GLASS CAST
AND
MOLDMAKING
Glass Fusing Book Three

Written by
BOYCE LUNDSTROM

Published by
VITREOUS PUBLICATIONS

Published by Vitreous Group/Camp Colton
© 1989 Vitreous Publications

All rights reserved. No part of this publication may be reproduced, stored in a retrieval system, or transmitted, in any form or by any means, electronic, mechanical, photocopying, recording or otherwise, without the prior permission of the publisher and the copyright holder.

Library of Congress Catalog Card Number 83-50657

ISBN Number 0-9612282-2-9

Edited by Kathleen Lundstrom

Designed by Linda Andrews

Photography by Joan Malone, Linda Andrews and Boyce Lundstrom.

(Title Page) Pate de verre "Pansy Dish", 8" x 8", Michael Barton.

Introduction

During the past year and a half, the two most asked about and talked about subjects among my students, and others expressing interest in becoming students, have been pate de verre and casting. It is almost as if these two glass processes had just been invented and the word is getting around. Yet these two methods of working glass are older than glass blowing and date back to 300 B.C. Manufacturers and suppliers are throwing these "buzz words" around a lot too. I am asked by a supplier if pate de verre can be fired in a top fired fusing kiln or is it necessary to use a side fire one, or whether I know of a casting material that will take higher temperatures than plaster-silica mix.

The processes of fuse forming glass cullet, frit, powder, or molten glass, into molds will be referred to, in this volume, by the process name most generally used by my peers in the glass art world. These terms are, in some cases, less than ideally descriptive, and that has been frustrating because it is my job, as author, to convey information as clearly as possible. However, historical use has won out, since I do not want to confuse the issue by inaugurating new terminology. Lost wax casting refers to the part of the moldmaking process used to create a cavity into which can be poured one of a variety of materials, from plastic to bronze, as well as glass. Pate de verre, paste of glass, refers to a method of using very fine particles of glass packed into a mold and then fuse fired. Crucible casting has nothing to do with casting crucibles, but describes the container that holds the molten glass before it is cast. The chapter on crucible casting could have been called crucible melting, but then we don't melt crucibles either.

In the first chapter of this book I do try to distinguish between pate de verre and frit casting, probably because pate de verre has such an historically recognizable look. I believe that historical references to processes are important. Over time, when a new look has evolved, due to a process or technique, the art historian will change the terminology accordingly to distinguish that particular type of work. Having an historical reference to glass work often makes the work more enjoyable, and gives the artist-craftsman a basis for comparison and thought.

It is my hope that you will find some of the answers you have been seeking about casting methods and mold materials in this book. I wanted to provide the most up-to-date information possible on the subjects covered, and I believe we have succeeded, in general. It seemed like a losing battle at times, when the activity surrounding the documentation of a process for this book led to new discoveries. It became hard to know where to stop so that the book could go to the printer! As always, our enthusiasm for glass working and our attitude that if you wonder about it, "try it", led to new mold formulas, new equipment and variations on processes.

Sometimes the simplest ideas come from the frustration of a glass process not working or from working in a studio when a variety of processes are being carried out at one time. While Mike Barton and I were working on pate de verre, I became very frustrated at the time it took to pack a mold with powdered glass, then fire it, then pack it some more, then fire it again, then pack the inside with mold material before the final firing. There was a limit to the time we could spend documenting that one process, and I had the feeling that I wouldn't be able to finish one piece. During one of our brainstorm sessions, two days before the end of our planned time together, I suggested a new method for doing pate de verre vessels without an inner mold and with only one firing. Spin-casting pate de verre, using centrifugal force to hold the particles of glass against the side walls of the mold, was the suggested solution.

The whole crew said "try it", so Peter Wendel, Mike Malone, and I gathered discarded parts of various equipment from around the studio and that evening we put together what I believe to be the first spin-cast pate de verre machine. The next day Peter and Mike Barton made a large mold, applied glass paste to the inside wall, and inaugurated the spin caster. Yes, it worked—sort of. And no, you won't find spin-cast pate de verre in this book; the process needs more development.

I believe it is a good policy not to publish information until it has been tested over and over. Often something that worked once or twice fails with the next two tries; fifty percent success is not enough to recommend a method. Experimentation based on information assimilated in our initial test studio work continued separately in the studios of each of us as the months passed. As the information base continued to expand, I began to realize how preposterous it was to have ever thought that I could have "final" results to publish. As you absorb the contents of this book, you too will see the never-ending potential for development of additional kiln forming glass processes.

It is my hope that you can use the basic information offered here as a "jumping off" point for developing a variety of techniques and combinations of techniques. As you use this book, you'll be a participant in causing it to become outdated, for as things are tried, new things are discovered, new materials invented, and so...

Table of Contents

INTRODUCTION		iii
ACKNOWLEDGEMENTS		5
CHAPTER 1	PATE DE VERRE	9
	The Process	
	The Molds	
	Preparing the Glass	
CHAPTER 2	LOST WAX CASTING	19
	Wax Models	
	Investing the Wax	
	Filling the Mold Cavity	
CHAPTER 3	CRUCIBLE CASTING	27
	The Crucible Furnace	
	Studio & Equipment	
	Melting Cullet	
	Sheet Casting	
	Coloring Float Glass	
CHAPTER 4	CASTABLE MOLD MATERIALS	45
	Binders, Refractories, Modifiers	
	Analyzing Castable Recipes	
CHAPTER 5	MOLD FORMULAS	57
	Recipes	
	Equipment	
	Mold Boxes	
	Models, Patterns & Extraction	
CHAPTER 6	METAL, CLAY & SAND MOLDS	71
	Steel Molds, Cast Iron & Brass	
	Clay Molds & Models	
	Sand Molds for Casting & Kiln Forming	
CHAPTER 7	FIBER MOLDS	91
	Forming & Firing Fiber	
	Applications	
	Zircar Fiber Ceramics	
CHAPTER 8	SAFETY	107
	Hazardous Products	
	Safety Equipment Sources	
CHAPTER 9	TECHNICAL INFORMATION	113
	Supply Sources	
	Annealing Graphs	
	Temperature Conversion Chart	
	Firing Schedule for Duplication	
	Technical Aspects of Fusing with Bullseye Glass	
	Annotated Resource Index	
EDUCATION AT CAMP COLTON		136
ABOUT THE AUTHOR		138

I have been very fortunate in having a wonderful group of past students and teaching associates work with me on this book. We gathered together for a few wintry weeks at Camp Colton and punished the studio, attempting to re-demonstrate everything I would write about. You will see their sometimes smiling, sometimes goofy, always dedicated faces, as you progress through these pages. And you will see their art.

Ruth Brockmann, Dan Ott, and Peter Wendel have taught with me for many years. They are excellent teachers, as well as fine artists, and they know the material glass thoroughly. They helped so much by being great communicators. Ruth maintains her glass studio in Seattle, Washington, Peter his in Corvallis, Oregon, and Dan is presently living and working full time at Camp Colton.

Doug Pomeroy has been a teaching assistant at Camp Colton and has a voracious appetite for experimentation. Doug lives for glass and is one of the quickest studies I've ever known of any new process. He maintains his studio in Corvallis, Oregon.

Mike Dupille, Seattle, Washington and Linda Andrews, Kennewick, Washington, both come from a commercial graphic art background, and both have recently been infected with glass madness. Linda helped with the photography, has redone all of the illustrations, and generously agreed to do the layout art and design for the book. Both Mike and Linda have great talent and will undoubtedly make glass work their entire livelihood in the near future.

Michael Barton, whose art work in glass has always been among of the most unique and most unsung of that of any of my art glass friends, is a studio commission artist in Sacramento, California. Michael's quiet participation and unique vision was a refreshing contribution.

Joan Malone, who is newly trying her hand at glass art, provided much of the photography, while Mike Malone gave his time, tools, and knowhow in preparing and maintaining experimental equipment. Joan and Mike have a glass studio in Bellingham, Washington.

So here is a special thanks and an open invitation to come back when the snow falls and the icicles grow on the steep banks of Canyon Creek to:

Ruth Brockmann	Peter Wendel
Michael Barton	Dan Ott
Linda Andrews	Mike Dupille
Joan Malone	Doug Pomeroy
Mike Malone	

Acknowledgements

Others who helped did not join our wonderful, whacky, winter session. Robert "Do-glass" Ross constructed the first prototype glory hole, as well as drawing the initial breakaway illustrations. Dan Schwoerer who, as always, lent his support for more information for fusers, provided the technical paper on use of Bullseye glass. Many others sent me up-to-date information on how the processes they had learned at Camp Colton were working in their personal studios. And many artists provided photos of their work, included herein. Thank you all.

A special thanks to my wife, Kathy, who pummelled me until the job was done well, or corrected my mistakes and said little. Kathy made this book readable; she found form in chaos, and she keyboarded 100% of the material into the Mac.

Thanks a thousand times everyone and thanks a million Kathy!

CHAPTER ONE
Pate de Verre

Pate de verre is finely ground powdered glass, made into a paste by mixing with water and a binder. This paste is applied to the inside of a mold using a brush or pallet knife or by dabbing it on the surface with fingers. The finely ground glass can be made from colored glass or it can be clear glass colored with metal oxides or enamels. It is often subtly color blended as it is packed into the mold in thin layers, resulting in sensitive shading. Pate de verre does not flow when fused. Slow, soak firing develops compaction of the fine grains into a solid object, which develops the rich "crystalline" quality of alabaster or jade.

The plaster-composition molds used for pate de verre crumble away from the finished object after firing. The surface of the finished work is usually matte, and the glass shines with an inner light, due to the low-temperature fusing of the small grains of glass. The outer surface of the pate de verre object often needs reworking, such as polishing or acid etching, to bring out the subtle characteristics of the glass beneath. Pate de verre can be made into a hollow vessel, or it can be cast solid. The thickness doesn't make an object pate de verre; the size of the particles and the light refractive quality does.

Placing frit or cullet into a mold and fusing it together is not always pate de verre. In the pate de verre process, the finely ground glass is packed into the mold, specifically placing colors in design areas, building up layers of different colors, one behind the other, to create shading and depth of color. This technique was developed by French artists at the turn of the century, during the art nouveau period. I believe that craftsmen who pack molds with large pieces of frit, large enough that individual "pixels" of frit can be seen in the final piece, are not doing pate de verre. I would rather call this adaptation of the French process frit casting.

The artist who sets out to accomplish the technique of true pate de verre needs four things, a suitable mold material, finely ground glass, a kiln with a controller, and the persistence to carry out the packing of the mold. The first three can be learned by reading this book, but the last requires patience from the individual hoping to prevail using this technique. Packing a mold, then firing it two, and possibly three, times is a solitary process.

Pate de verre bowl.

THE PROCESS

There are two basic processes used for packing and firing pate de verre molds. In addition, there are many variations of each step in these methods. The first is described as the single containment mold process, fired with the top open. The glass paste is packed into a mold, either as a thin wall or packed solid. The second method uses an inner and an outer mold to contain the glass on both surfaces.

In the single mold process, a pattern is made of wax or clay, and then a mold is made with a suitable mold material. I suggest Mold Mix 522 or N&N 965 (see Castable Mold Formulas). It may be necessary for the mold containing the glass to be fired two or three times, so it is important to make a sturdy mold, with even wall thickness of approximately two inches. The mold should have an opening large enough to allow access to all surfaces. If the opening is too small, it may be necessary to make this outer mold in two parts, cementing the mold together with fiber glue, after packing the glass, but before firing.

Opposite page: "Pity and Compassion", "Pansy Dish", "Tommy", pate de verre, Michael Barton.

Mike Barton inspects pate de verre work fired to 1400°F without an inner mold.

MAKING AN INNER MOLD

CLAY PATTERN

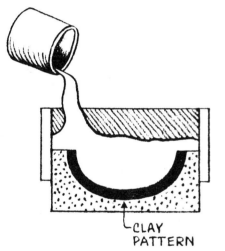
CLAY PATTERN

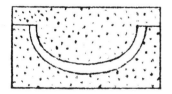

The glass paste is applied in thin layers to incised areas, or where design dictates, eventually covering the entire inside surface of the mold approximately 1/16" thick. This first layer can be air dried or dried with a hot air gun. A second layer (1/16" - 1/8") can then be added over the first, creating depth of color and design. After applying the second layer of paste, the molded object is fired to 100°F below the flow point of the glass. For

Bullseye glass this is 1300°F. The firing is held at this temperature for three to four hours, until the entire mold has soaked to an even temperature. After cooling, the glass will have lost approximately 30% of its original volume.

Additional layers of clear or colored glass are applied to the mold, until the added glass equals approximately the total of the first two layers in volume, before the piece is re-fired. The second firing should be done just as the first one was, soaking the glass and mold at 1300°F. After it is heat soaked for three to four hours, the object can be fired to maturity. Or it can be cooled, so that the intended void inside (in the case of a vessel) can be packed with mold material to contain it. In the case of bowls or other gently sloping objects, once the glass has fused, the temperature *can* be then raised approximately 100°F until the exposed surface is shiny.

In the second pate de verre method, an inner mold is made to hold the glass against the outer walls. This can be accomplished in two different ways. The first is by making both the inner and the outer molds from an original pattern. The second is by packing the inside of a piece with suitable material after the consolidation firing, but before the final firing.

The original clay pattern is placed in a mold box with two inches of space around all sides, cast with a suitable mold material, after applying a

Pate de Verre

parting agent, and allowed to set. The mold and pattern are inverted and the mold box raised halfway up the side of the mold, extending the box edge to slightly above the top of the mold. After making alignment notches on the first plaster cast, and applying a parting agent to the surface of the mold that will come in contact with the inside mold (as well as to the pattern), cast the inside of the form. The two-part mold is allowed to cure for two hours, then heated in the kiln for 3 to 5 hours at 200°F. After drying, the two part mold will separate easily. The clay pattern is removed in pieces or the entire mold may be placed in water to remove clay that does not part easily.

Packing the outer mold with glass paste and firing once, without the inner mold in place, to consolidate the first two applications of glass is a matter of individual technique. It is important to add enough extra volume in creating the outer walls that, when the inner mold is in place, it will be elevated off of the outer mold. Weight the inner mold with a brick. As the glass shrinks during the final firing, the weighted inner mold will sink into place, keeping the glass pushed against the outer walls.

For vertical-walled vases or forms with reverse curves, the process is begun as it is in the two-part method. The mold is packed in layers, and the glass consolidated by soak firing below the flow point of the glass. Then the inside of the mold is packed with pre-fired (used) mold material and chopped fiber, gently compressed against the sides as the cavity is filled. It is then ready to be fired to maturity. This process is necessary to keep design detail in the proper place and to keep the glass from flowing to the bottom of the mold. If the glass paste has good packing properties and is consolidated by at least one firing at a temperature lower than that at which the glass would mature, very fine detail can be maintained on vertical-walled objects.

THE CASE FOR USING LOW-MELTING-TEMPERATURE LEAD GLASS FOR PATE DE VERRE

In the past, it was very difficult to use glass frit that matured at over 1400°F for pate de verre, because mold systems made of plaster compositions lose most of their strength at 1350°F -1400°F. The addition of binders that hold the refractory particles together at higher temperatures (such as calcium alumina cement) results in a mold that does not readily separate from the glass surface. Mold systems that have more than 50% plaster shrink excessively over 1400°F and cause cracks in the mold, which cause unwanted lines in the finished object.

Using a lead glass with a low melting temperature, is one way around having to extend the bonding properties of plaster composition molds. However, the low viscosity of lead glasses creates a touchy firing situation. A variation in firing temperature of fifteen to twenty-five degrees can cause images to move and color mixing that may not be wanted. A 20% to 25% lead glass mixed with 80 to 120 mesh enamels seems to be a very satisfactory solution to using most plaster-bonded mold materials. Most high-lead glasses will consolidate around 1050°F-1100°F and mature between 1300°F and 1350°F. These temperatures are well within the stability ranges of many plaster composition investments.

Packing the inside of a vertical wall with fiber and ludo after consolidating the glass.

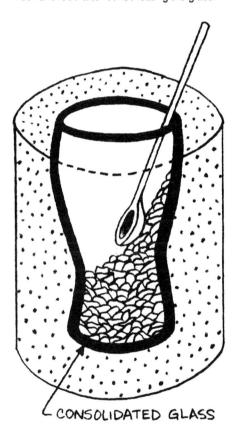

CONSOLIDATED GLASS

USED MOLD MATERIAL & FIBER BEING COMPRESSED INTO PLACE WITH A SPOON.

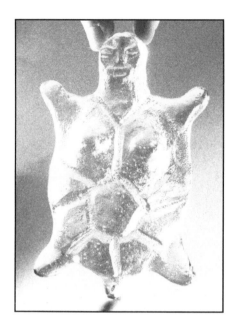

Frit cast turtle woman with copper fish inclusion.

Spooning frit and cullet into two molds. Note the copper fish placed on the middle layer of the turtle mold.

Pate de verre "Rodeo Fish" by Boyce Lundstrom.

Clear lead glass powders can be mixed with transparent enamels to impart color. The addition of 10% to 20% enamel, by volume, is sufficient to create a dense enough color for most pate de verre. Adjusting the coefficient of expansion of the enamel to suit the base glass is important when making large or thick pieces.

The colored glass powder is mixed with a water and gum arabic solution to make it into a paste. Twenty parts hot water to one part gum, mixed well, can be added to small amounts of glass powder and blended on a glass plate with a pallet knife. This mixture will harden upon drying, so should be mixed as needed. Whether using enamels for coloring or using colored frit, the fusing process is the same. However, enamels work best with low-melting-temperature lead glasses, because their flex points are close.

THE CASE FOR USING BULLSEYE

I enjoy using Bullseye frit, because the pallet is a good one and I have a lot of scrap leftover from fusing classes. The development of Mold Mix 522 was a response to wanting to use Bullseye glass for pate de verre and frit casting. During the experimental phase of trying to understand the pate de

verre technique, and attempting to shorten the multiple firing process, we fired many bowl forms using Bullseye frit in open-topped, 522 molds. We applied the glass paste, mixed with gum arabic solution, in successive layers to a mold no more than two hours old, occasionally applying heat with a hot

Pate de Verre

air gun to set the glass. As soon as all the glass was in place, we fired the 522 mold and the glass in a *top-fired* fusing kiln. The firing method was to raise the temperature to 1400°F over a five hour period, then hold for approximately one hour, visually checking the surface flow. The glass matured around the lip of the bowl first, then slowly matured down the walls to the bottom of the bowl. The more elaborate the incised design on the mold surface, the less the glass slipped down the walls of the mold. Many pieces were very successful, especially those with approximately 3/16" of applied paste.

Further experiments were carried out mixing G.N.A. frit with Bullseye frit. G.N.A. has the same coefficient of expansion as Bullseye, but melts (flows) at a temperature approximately 70°F higher. By mixing G.N.A. frit particles of between 30 to 50 mesh and Bullseye frit, 50 mesh and finer, we found that the stiffer G.N.A. held the glass up on the walls while the Bullseye fluxed and matured. This mixing of large, high-melting-temperature particles and smaller, lower-melting-temperature particles is similar in concept to mixing enamels with lead glasses.

Pate de verre mask by Ruth Brockman.

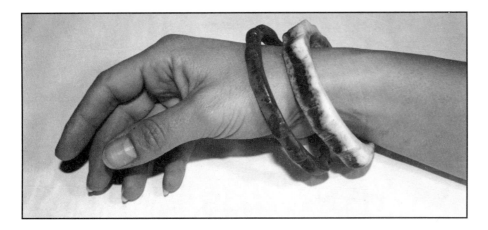

PREPARING GLASS FOR PATE DE VERRE AND FRIT CASTING

It is necessary to have finely ground glass for pate de verre, to achieve the diffused light quality or inner glow for which the glass process is known. If the individual pixels of frit can be seen in the finished fusing, the glass particles are too large. Densely packed glass particles contain a minimum of air in the paste, so that there is a minimum of shrinkage, and less movement that could cause the design to slip from incised areas of the mold during melting. Using a mixture of varied grain sizes enhances the packing quality of the frit, and makes a more plastic paste that will adhere better to the mold surface.

Different crushing or milling methods create different assortments of grain size. Also, different glasses crush differently using the same crushing machine. It is important to understand that pate de verre glasses can be made to apply and melt evenly, if the proper size of glass frit is used. There are many ways to crush and sort glass in the studio.

Portrait of a boy looking through a window, by Mike Barton.

Pate de verre and cullet mask by Ruth Brockmann.

Making fractured frit by dropping hot glass in a water bucket.

CRUSHING METHODS

Metal tube and iron bar crusher	Coffee or grain mill
Frit by heating and quenching in water	Commercial roll crusher
Garbage disposal (hammer-mill type)	Limb mulcher (fly-wheel type)
Mortar and pestle (porcelain)	Hammer mill (with 1/4" screen)

All these crushing methods except those using a commercial roll crusher or hammer mill are relatively inexpensive and can be accomplished in most studios. Fritting by quenching in water and hand grinding with a mortar and pestle are the only ways to be sure that small amounts of iron and stainless steel are not left in the fine, crushed glass. When bits of metal are left in the glass from a crushing method, they can be removed with a magnet. Good magnets can be had by disassembling broken audio speakers. Auto parts stores and toy stores sell magnets, but they are usually not as powerful as those found in large speakers.

I have tried all methods of crushing glass to a fine powder, except the commercial roll crusher. I wish I had one! We presently use three methods in the teaching studio. In one method, we put large pieces of glass in a stainless steel mixing bowl and heat it in the kiln to 900°F. The bowl is removed and the glass is dumped into a bucket of water. This fractures the glass into pieces approximately 5 to 10 mesh. This is a good size for filling open faced molds or for loading a flask when doing lost wax casting. However, 10 mesh is not a powder and needs to be crushed finer to make a paste or blend colors without seeing the individual frit pieces. It is harder to crush quenched frit (hot glass fritted in water) than to hammer crushed frit, because the quenching has tempered the glass. Yet this method makes the cleanest frit, free of any metal contamination.

A second method is grinding with a hand-held pestle in a porcelain mortar. A handkerchief, folded and placed in the palm of the hand, helps cushion the friction of the pestle. Approximately one inch of frit is placed in

Pate de Verre

the mortar, rested on a solid, low table. Pressing very hard, the pestle is rocked from side to side five or six times, then stirred and rocked again. In about 10 minutes a strong, diligent person can make six ounces of powdered glass.

Most of the glass we crush at Camp Colton for pate de verre is done in a garbage disposal. I have mounted the disposal in the lid of a 30-gallon, metal drum. A hole has been cut in the side of the drum to fit the hose of a 2 1/2 hp. shop vacuum. The glass is initially broken, by hitting it with a heavy metal pipe. Wearing eye protection, gloves, and a respirator and, with the vacuum running to remove the glass dust, I start the disposal and drop in pieces of glass. After grinding for ten to fifteen minutes, it is necessary to give the disposal a rest, so it doesn't overheat.

After turning the disposal off, I wait twenty seconds while the vacuum removes the remaining glass dust, and then remove the vacuum hose from the side of the drum and put it into the disposal to remove any chunks that have been rounded off and will not grind up. This prevents the pieces from wedging between the hammers and the side slots, which could keep the disposal from re-starting. If glass does become jammed, I have found it easily removed by sticking a large, flat-head screwdriver into the disposal exit port and prying against the underside of the hammer plate.

I use a 1/2 hp. hammer mill garbage disposal. A hammer mill disposal has two or three stainless steel swing arms (knobs) attached to a rotating horizontal plate. These knobs can move back and forth on the pin, which helps keep the glass from getting wedged between the moving parts and the side cutting blades. Of course, I buy all my equipment from Sears. My first disposal crushed over 500 lbs. of glass before the side walls gave out. Not bad for $89.00 dollars!

Years ago, when I was blowing glass, I needed large quantities of bottle-glass frit to mix with batched materials for blowing. I rented a fly wheel type limb mulcher, and set it in the back of my pickup. At a recycling yard, I purchased two 55-gallon drums of broken, clear bottles. With the mulcher sitting on the tailgate and a piece of plywood across the sides of the pickup bed, I shoveled the glass into the mulcher, blowing it into the front of the pickup, under the plywood. After unloading the glass at my studio and

Removing the lid of a garbage disposal grinder to inspect frit.

Safety glasses and poke stick are essential to safety while grinding glass.

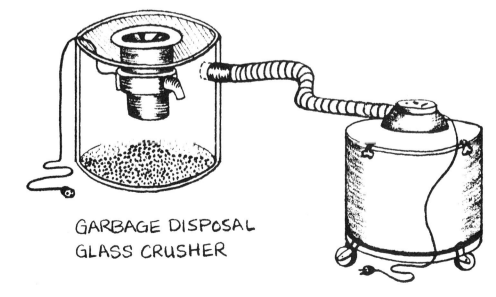

GARBAGE DISPOSAL GLASS CRUSHER

Pate de verre test samples demonstrate the results of various combinations of glasses for their ability to stay on a mold wall during the consolidation firing.

Sifted glass of various particle size.

hosing down the mulcher, I returned it to the rental yard. It took them a lot longer than I had thought it would to ask me what I was doing to the mulcher that made the blades so dull. Well, I owned up to crushing bottle glass, and from then on I paid them an extra $5.00 to sharpen the blades when I was finished. They actually appreciated the fact that the machine came back so clean!

After crushing in any mechanical grinder, the glass must be screened to various sizes, depending on the intended use. We have put together a set of screens that we purchased at a ceramic supply store. These screens fit inside of each other, forming an interlocking stack. The screens are 30, 40, 50, and 80 mesh. The glass is first screened through a window screen of approximately 17 mesh. I have found that the results from screening 10 pounds of garbage-disposal-milled glass are 6 pounds of material smaller than 17 mesh and 4 pounds larger. This proportion seems to hold true when the disposal is fairly new; as it wears out the proportions change. The larger frit contains no iron or stainless steel metal fines. This size frit is excellent for frit fuse casting and lost wax casting. The remaining finer frit is sorted by all screen sizes.

We did some experiments, proportionally mixing various sizes of frit to see which packed the tightest. Screen sorting was carried out for different

Pate de Verre

crushing methods, using the same glass. We placed 75 grams of various combinations of particle sizes of clear Bullseye (1101F) into a 3/4"-wide glass beaker 20" long, then vibrated the tube with a sander held against its side. This vibrating action packs the glass in a way very similar to adding a liquid. Since each mix had the same weight, the height of the glass in the tube represented its packing properties. The data from our tests follows:

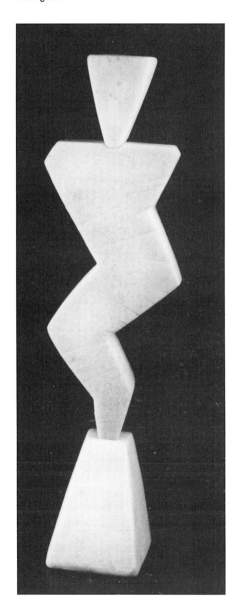

Pate de verre sculpture 2'6" high, by Mark Abildgaard.

SCREEN MESH SIZES

17-30	17-40	30-50	50-80	80-mill	LEVEL IN TUBE
				75g	20"
			75g		18 5/8"
		75g			18"
75g					17 1/4"
		25g	25g	25g	17"
	75g				16 5/8"
		37.5g	37.5g		16 3/8"
			37.5g	37.5g	16-1/4"
		37.5g		37.5g	15-3/4"
			37.5g		15 1/2"
	37.5g				15 1/2"
	40g	10g	10g	15g	14 1/4"
	40g	10g		25g	13 3/4"
45g				30g	13 1/4"
	50g			25g	13 1/4"

PROPORTIONS OF PARTICLE SIZE DISTRIBUTION DIRECT FROM GARBAGE DISPOSAL

17 Mesh to Mill	31g	16g	15g	14g	14 5/8"
30 Mesh to Mill		26.5g	25.5g	23g	17 1/8"

Various conclusions can be drawn from these test results, but perhaps the most important conclusion is that a *mixed particle size* will definitely pack tighter than an evenly graded particle size. Whether or not it is important to find the best mix for the tightest packing is a matter of personal preference for the look of the finished work. Packing properties affect how the paste applies and adheres to the sides of the mold and determine the amount of shrinkage the glass will undergo during fusing. Particle size determines the light refractive quality of the completed glass object.

Pate de verre is a French term for art glass work of a very specific nature, done by artists in the early 1900's. It is also a term for a technique of applying glass paste to a mold to achieve sensitive shading of hue and tone. When it is not done in the proper way, I don't believe it is pate de verre. Frit casting, lost wax casting and crucible casting offer glass qualities and techniques different from those of pate de verre. Perhaps these processes should be separated in terminology, so as not to confuse the solitary process and inward thought that is necessary to achieve good pate de verre with the more spontaneous processes and results that are achieved using other casting methods.

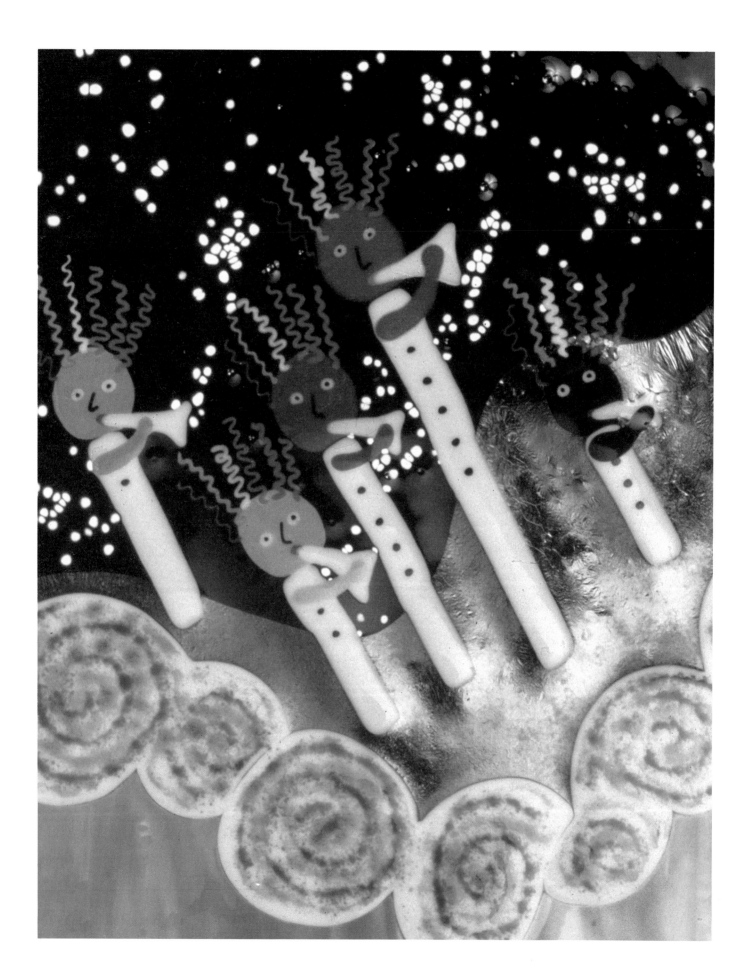

CHAPTER TWO
Lost Wax Casting

Making a wax model, investing it in a plaster composition mold, then melting and vaporizing the wax from the mold to create a void the shape of the original model, is known as lost wax. After the wax is removed, there are three options for filling an interior mold cavity with glass. These procedures and the temperatures necessary differ quite a lot from those used for open face mold casting. They are as follows:

1. Fill the mold cavity with crushed or ground glass, placing extra glass in the flask and sprue (formed into the original mold). This type of mold is usually fired from 1500°-1550°F. The excess glass contained in the built-in flask flows into the mold cavity as the glass melts and consolidates.

2. Support a clay crucible with a hole in the bottom, containing glass, on top of the mold and fire the mold, crucible, and glass to 1650°F. The glass will melt and run out the hole in the bottom of the crucible, thus filling the mold cavity.

Cross-section of a lost wax mold showing the finished glass piece before it is removed from the mold.

3. Fire the lost wax mold to 1400°F in one kiln and melt glass at 2000°F in a ceramic crucible in a separate kiln. Pour the liquid glass into the hot mold and hold the kiln at 1400°F for one hour to allow the glass to fill detailed areas.

Opposite page: Detail, "Harmonic Convergence," Richard LaLonde. This is not lost wax.

Organic objects may also be invested in a plaster mold.

Double boiler set up for melting wax.

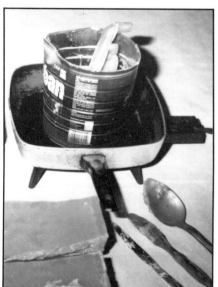

These three processes produce different surface qualities and light qualities within the glass. The types of glasses used will also effect the choice of casting technique. The first process is best suited for low flow temperature glasses such as lead crystal. Since the glass is contained in the mold, if the glass is not pliable or moving as the mold reaches 1350°F to 1400°F, (at which time the plaster composition mold shrinks dramatically), then the mold will crack due to the fact that the glass does not move as the mold shrinks around the glass. Mold Mix 522 is suggested for this procedure.

When using the second process, the glass does not enter the mold until the mold has already shrunk. Therefore, the only requirement is to not exceed the temperature of the mold material. R&R 965 is very stable up to 1700°F, so would be an excellent castable for this process.

In the third process, the glass is heated separately from the mold, therefore the mold does not have to be heated past the temperature where plaster composition molds shrink excessively and lose strength. If failure does occur, it is usually due to thermal shock of the mold material. Mold mix 50-50 works well for this process.

Wax is not the only material that can be melted or "burned" out of a mold. Other organic materials may also be invested and "burned" or vaporized out of a plaster composition mold. Woody materials such as pine cones, woven straw baskets, or sticks will burn as the mold reaches 600°- 800°F. As the mold reaches 1200°- 1300°F all that is left is a very small amount of ash which does not cause adverse effects to any glasses except high lead glasses. Therefore "lost wax casting" can be expanded into lost stick or lost apple casting if you wish.

The type of wax used for making original models is a matter of individual preference. Micro crystalline wax is a commonly used wax and is available from many hobby stores. Micro crystalline wax is a medium-hard wax that becomes very pliable at 90°F or when worked in the hands. It can be carved to fine detail, after placing in the refrigerator to cool to 40°F. For very fine detail, many hard waxes are available through dental supply stores. A good substitute wax for micro crystalline wax is a combination of paraffin and beeswax. Between 10 to 20 percent beeswax added to paraffin makes a very suitable wax for hand modeling.

Melting, then forming, the wax into various shapes requires the use of a hot plate, a double boiler, or an electric frying pan. Wax should not be heated beyond 212°F; therefore the melting container should always be surrounded by water. *Wax that becomes too hot will vaporize creating an explosive situation.*

Forming the wax into sheets is often the first step in forming original wax models. To do this, a plaster slab is made, the size of the wax sheets to be formed. After allowing the plaster to cure, soak the plaster slab in water, build a dam, and pour the wax over the wet plaster. After cooling, the wax will separate from the plaster. Wax sheets can be heated in the sun or in a warm oven to make them very pliable.

Three dimensional objects, vases, and bowls can be formed by pouring hot wax into wet clay forms. Clay is formed in any of the various ways into a hollow form. After drying for five or six hours, the clay becomes hard enough

Lost Wax Casting

to handle and carve. Fine detail may be carved into the interior of the clay vessel. The wax is then poured into the cavity and the vessel is turned or rotated until the wax covers the surface. The excess is poured back into the melting pot. Six to ten applications of wax applied in this fashion, allowing one minute between hot wax pours, will form a one quarter inch thick vessel wall. After the wax has cooled, the clay is easily peeled off the outside. If the clay gets too hard for easy removal soak the entire piece in water.

After the wax is complete, a wax sprue and flask should be added before the wax is invested in the plaster mold material. A sprue is the connection between the wax object and the flask. It may be large or small depending on the size of the object to which it is attached. The sprue should

Peter painting the plaster splash coat of a fish napkin ring.

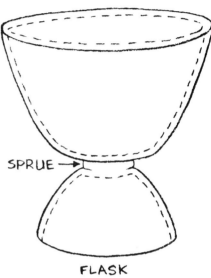

Pouring the second part of a two-part master mold. Multiple waxes were taken from this mold.

be at least 3/4 of an inch across to facilitate the flow of glass into the mold cavity. After the glass is cast into the mold and annealed, the object will be separated from the flask by cutting off the sprue. The flask is like an inverted cup and serves the purpose of holding additional glass that is needed to fill the void after the glass melts and consolidates. It is not always necessary to have a sprue. The sprue, as an addition to the wax object, allows even mold wall thickness all around the wax model.

Lost Wax Casting

INVESTING

When the wax model is complete with sprue and flask, attach the bottom of the flask to a glass plate very securely by heating the glass and the wax with a hand-held torch. Build a box around the wax, leaving approximately 2" of space on all sides. Apply a suitable wetting agent to the wax to insure the investment material will closely contact the wax object. A liquid cleaner, such as Murphy's Oil Soap, (mixed 50/50 with water) or WD-40 work well. Invest the wax by pouring a suitable mold material into the secured mold box. Because most wax models contain undercuts, or are forms that may trap air pockets in places that cannot be seen from the top of the containment mold box, it is often necessary to vibrate the mold box while the investment material is still very liquid. This will allow bubbles next to the wax surface to rise. Very intricate designs should be coated with a splash coat of plaster investment before they are placed inside the mold box. Mix a small amount of the investment material to be used and cover the wax model, using a brush or by splashing the investment onto the wax. This will harden before the rest of the investment is cast into the mold box. With most investments this splash coat will adhere without ill affect.

Allow the mold investment to cure for the time suggested for the particular material. After it has cured, elevate the mold with the flask side down over a pan of water in your oven or kiln. Heat the kiln to approximately 250°F; the steam will rise, causing the wax to melt and run into the pan of water. Add water as necessary, until all the wax has run from the mold. Not all of the wax will be removed with this process; after the wax has quit flowing there is still wax contained within the investment material. Remove the mold and water pan from the kiln and, after cooling and inverting the mold, place it back in the kiln on stilts and fire it slowly to 1200°F, with the door slightly cracked. This will vaporize the remaining wax. Expect some

Mold set up ready for pouring.

A series of wax fish in three stages: partially bent wax fish, fish mounted on sprue, fish with a plaster splash coat.

Applying WD-40 as a separator to a glass containment box, held together with plasticine clay.

Applying the plaster splash coat after the containment mold has been attached to a base.

A wax model made by pouring wax into a leather hard clay mold with an incised design on the inside.

smoke and a lot of stink; *be sure to have adequate ventilation.* If there is a lot of smoke, all the wax may not have run out of the mold. In this case *do not* open the kiln, the addition of fresh air may cause the wax vapors to catch on fire. Turn the kiln off, let it cool, then start over.

After the mold has reached 1200°F, all of the wax will be vaporized and any of the three methods mentioned earlier may be used to fill the mold cavity with glass. In the case of the second or third method, it is not necessary to cool the mold. If glass has been heated to 1000°F in a crucible with a hole in the bottom, in a second kiln, the crucible can be placed on top of the mold and then fired to casting temperature, or glass can be fully melted in a second kiln and poured into the hot mold.

Cool molds slowly, even when they don't contain glass. Plaster composition molds thermal shock due to the large volume change upon cooling or heating. Molds should be elevated off the kiln shelf, and placed in the center of the kiln whenever possible. These methods will insure even heating and cooling.

When annealing glass in a mold, the thickness and evenness of the mold wall must be considered. Uneven wall thickness can cause uneven cooling. Long soak time at annealing temperatures is necessary to insure the glass gets to the annealing temperature. It takes approximately three hours soaking time for two inches of plaster composition mold material to equalize the thermal gradient caused by cooling. Therefore, cool the kiln to

Lost Wax Casting

below the annealing point but above the strain point for two hours then turn the temperature up to the annealing point *to start* annealing.

Because castables are insulating, it will take extra time for the glass to cool to room temperature inside the mold. The mold may feel cool to the touch, but the glass may still be 300°F inside the mold. Remove the mold after you are sure the glass is at room temperature. The mold material should separate from the glass easily. A putty knife and a wire brush should remove 95% of the mold material. Washing the glass object in water will remove the rest of the plaster, but do not place the glass in water until it has been at room temperature for three or four hours.

The quality of the surface of the glass depends on the type of glass used in comparison with the type of mold material. In other words, some glasses work better with certain mold compositions. It is common to have a slightly dull or matte surface on the glass wherever the glass and mold have been in contact at temperatures over 1500°F. This surface can be acid etched or rubbed with pumice and water to bring the surface back to life. Acid etching can be dangerous; rubbing with pumice paste is only time consuming.

The excitement mounts as the mold is removed and the cast glass object becomes visible.

Wax face models in progress by Doug Pomeroy.

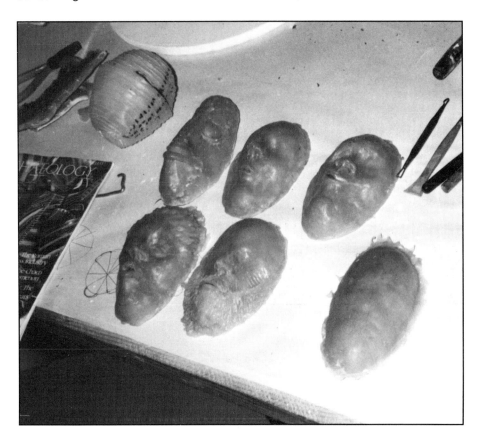

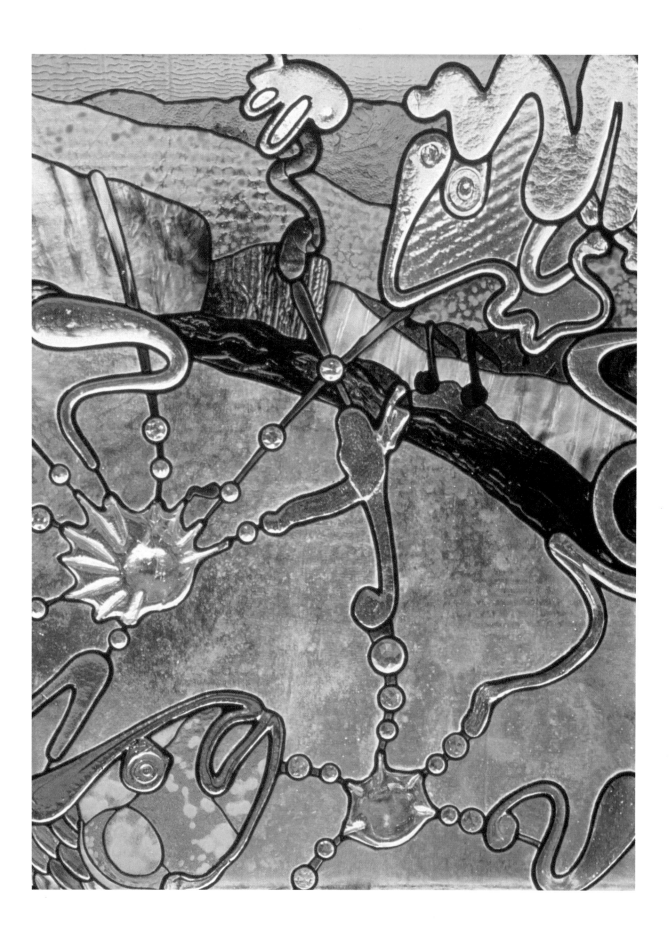

CHAPTER THREE
Crucible Casting

Once you have a crucible furnace set up in your studio, a new world of hot glass opens for you. The possibilities include pouring glass into molds, free forming lines and drips on the marver, fritting, making stringer and rods, making steam bubbles, blowing confetti glass or eggshell glass and, of course, offhand vessel blowing.

It's not necessary to have a large furnace to do any of these things, nor is it necessary to have a furnace that stays on for more than one working session. Crucible melting of scrap stained glass, float glass or bottle glass cullet can be accomplished in four or five hours from the time you turn on the furnace. A hot glass session can be as brief as filling one mold or casting two or three sheets. Pulling stringer, making shapes to use in future fusings, or blowing can use up any melted glass left in the crucible.

Gathering glass from the flue hole in the top of a crucible furnace.

Opposite page: "Fish Whistle", 28" x 40", using hot line elements, Boyce Lundstrom and Mike Barton.

Removing a crucible from an electric kiln, after turning off the power.

A crucible furnace is a kiln that contains one or more clay crucibles used for melting glass. Most side fired electric kilns can be used for melting glass in crucibles, but have limitations. It is difficult to reach and maintain high enough temperatures and easy to burn out elements. This is not to say that one cannot or should not melt glass in crucibles in an electric kiln. I have been using a Skutt Octagon 818 for three years with students, sometimes doing two melts in a day in four crucibles that each hold five pounds of glass. I have had to replace the elements after about 10 such firings, and the kiln bricks finally deteriorated to the point that the kiln needed a new lining.

Crucible Casting

Instead of rebuilding the octagon kiln, I removed the elements and lined it with 1 inch of 8-pound-density fiber blanket, set it up on a fire brick base, and added a propane burner. Now we melt approximately 20 lbs. of glass in one large crucible. A round hole was cut in the lid as an exhaust port and an opening from which to gather the glass.

There are many variations possible for kiln modifications. If gathering glass is going to be the main activity, raising the lid by adding a couple of courses of insulation brick, leaving a front opening, would be better than cutting a hole in the lid.

There are many different ways to construct a homemade propane burner, and diagrams are available in books on glass blowing. An idea of how such burners look can be gained by looking at equipment on the market that includes the type of burner you want. I presently prefer using a burner that doesn't need forced air. Natural draft venturi burners are available from Eclipse Fuel Engineering Co. and Maxon Burner Co. Many ceramic suppliers carry burners for small kilns.

A Sticktite flame retention nozzle on an 80,000 B.T.U. venturi burner is presently being used in my studio. This burner doesn't require a blower for forced air, so I don't have to worry about leaving the furnace on overnight or worry about the electricity going off, shutting off the blower and causing flames to rise 2 or 3 feet above the furnace.

For those who use forced air, there are many automatic shut-offs designed to shut the gas off when the power goes off. A drawback to this system is that, in favor of safety, the crucible will be lost if it cools long enough during a loss of power.

My students and I have made glass furnace burners from weed burners purchased from the local hardware store, used black pipe and vacuum blowers and are experimenting with hair dryers as the air source. These

Getting a good grip on the crucible before pouring glass into a mold.

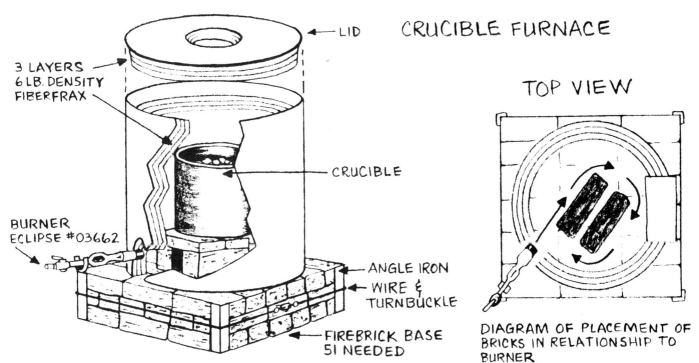

Filling a plaster composition mold by trailing glass from a punty rod.

forced air burners are unsafe if left unattended, but do get very hot very fast. They allow us to melt 20 to 50 lbs. of glass and cast it in one daily session. The pot furnaces are always empty and turned off at the end of the day.

You may wonder if gas can be dangerous or the recommended safety equipment a bit complicated and expensive. Why not use electric melting, which seems safe and easy to control, for melting glass? The reason is simple: quick melts (four hours) in small crucibles (approx. 5 lbs.) work well in an electric kiln, but larger amounts of glass melted for extended periods of time burn out elements.

The fluxes in the glass volatilize at elevated temperatures. These vapors attack the elements, causing them to burn out. This usually happens when the glass is still in the crucible. The higher the temperature, the quicker the elements burn out.

Although many ceramic kilns are rated at cone 10 (2350°F), they are not meant to sustain that temperature for long periods of time. There is nothing more frustrating than to have 30 lbs. of glass so stiff you can't get it out of the crucible, and there is no way to change the elements when the kiln is hot. You lose the hot glass session, and you generally lose the crucible. Gas furnaces are less expensive to maintain, they melt more glass faster and they reach higher temperatures.

THE STUDIO AND EQUIPMENT

Crucibles for melting glass are generally made of a high alumina or zirconia clay composition. This is because these materials are not readily fluxed by the molten glass.

Crucible Casting

Laclede-Christy, a clay products company, manufactures many standard shapes and sizes of crucibles made especially for glass. Their Mullac-VC is a 65% alumina, mullite-based crucible that I have found to work well for melting glass of any composition. Crucibles GC-28, GC-96, and GC-74, hold 5, 6 and 7 pounds, respectively, of molten glass. Many larger standard sizes are also available.

A very important aspect of any crucible is its shape, especially if it's to be removed from the crucible furnace and then returned, when empty, for annealing. Crucibles will thermal shock if not returned to the kiln while still red hot. Crucibles with slightly sloping sides and a large flat bottom are easiest to handle with crucible tongs and are not likely to tip, knocking other crucibles over, when returned to the furnace. Rule of thumb: the width of a crucible should be two thirds its height.

Making your own crucibles allows great freedom to experiment, is less expensive than buying commercially made ones, and is relatively simple. One of three forming methods is generally employed: throwing on the potter's wheel, hand building, or slip casting. The following recipes may be used for any of the three forming methods. Ingredients are given in parts by weight, and the total alumina and silica (refractories) indicated in the chemical statement beneath the title.

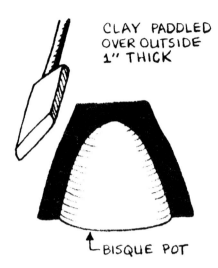

CLAY PADDLED OVER OUTSIDE 1" THICK

BISQUE POT

THE CLAY WILL LIFT OFF THE BISQUE POT AFTER DRYING 2 OR 3 HOURS.

CBL-3 (52% Al_2O_3 - 32% SiO_2)
28% Calcined Alumina
   (alumina oxide)
8% Kaolin E.P.K.
25% Grog 420
10% Ball Clay OM-4
28% Fire Clay Lincoln
1.5% Bentonite

CBL-4 (55% Al_2O_3 - 35% SiO_2)
35% Calcined Alumina
   (alumina oxide)
35% Ball Clay
20% Grog 420
10% Pyrophyllite

For hand building or throwing, mix ingredients dry, add water to the clay mix, and wedge until an even consistency is achieved. If the clay mix is too wet, spread it out on canvas or a plaster bat and air dry. Slip casting requires a different treatment of this same mixture, as outlined in ceramics manuals.

Crucible formula clay bodies are not very plastic, but can be thrown on a potter's wheel keeping the walls and bottom at least one-half inch thick. When hand building over a plaster form, let air dry five hours or until leather hard, then paddle outside surface to increase density. Form walls at least one-half inch thick. Air dry four or five days, then fire to cone 10. Crucibles made to hold more than 10 lbs. of glass should have thicker walls.

Any pot made of white stoneware clay and fired to an 1800°F bisque may be used as a glass melting crucible. These porous crucibles may last three or four firings before cracking. Whereas bisque stoneware may not be as long lasting as high fire clay formulas, they can generally be acquired from any professional potter.

When working with glass from a crucible, it is necessary to have an annealing oven to anneal most of the pieces you might make, although stringer, small rod, eggshell-thin confetti glass and small pressings can be air annealed or covered with fiber blanket for an adequate anneal. Heat proof gloves, shears, a hand torch, safety glasses and crucible tongs are all necessary items when working with molten glass. A glory hole, marving table, and various blowpipes and punty rods are usually added.

A layout for studio equipment should allow the work process to flow smoothly. Ways of achieving this are more readily learned by working with hot glass, than by reading a description. Keep in mind, as you organize your studio, that most hot glass work is accomplished best with a partner or team. Working with a partner becomes a dance; the more you work with one person, and the better you know him, the closer you can dance without stepping on toes.

Organize the studio around the process to take place, not around a central piece of equipment. Try to choreograph the moves you think will be involved in the work session, so that you can anticipate how equipment placement will work for you or against you. For example, when pour casting from a crucible furnace, the lid of the furnace must be removed to get at the crucibles. Leave enough room to get to all sides of the crucible furnace and plan a place to set the hot lid while removing the crucible.

Crucible Casting

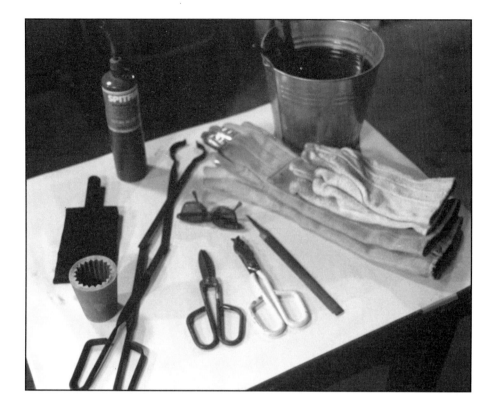

Tools used for casting and pressing; good gloves and eye protection are a must.

The hot shop at Camp Colton set up for casting and glory hole work.

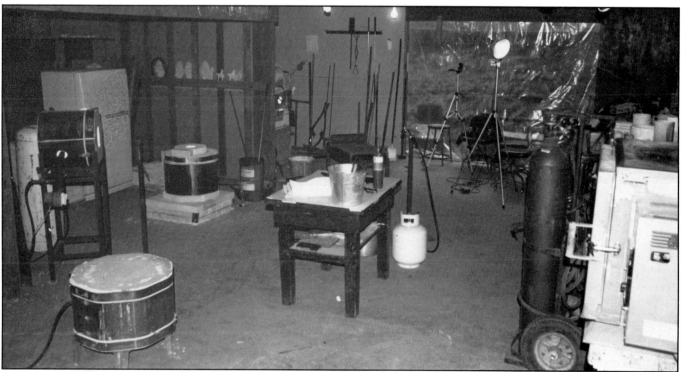

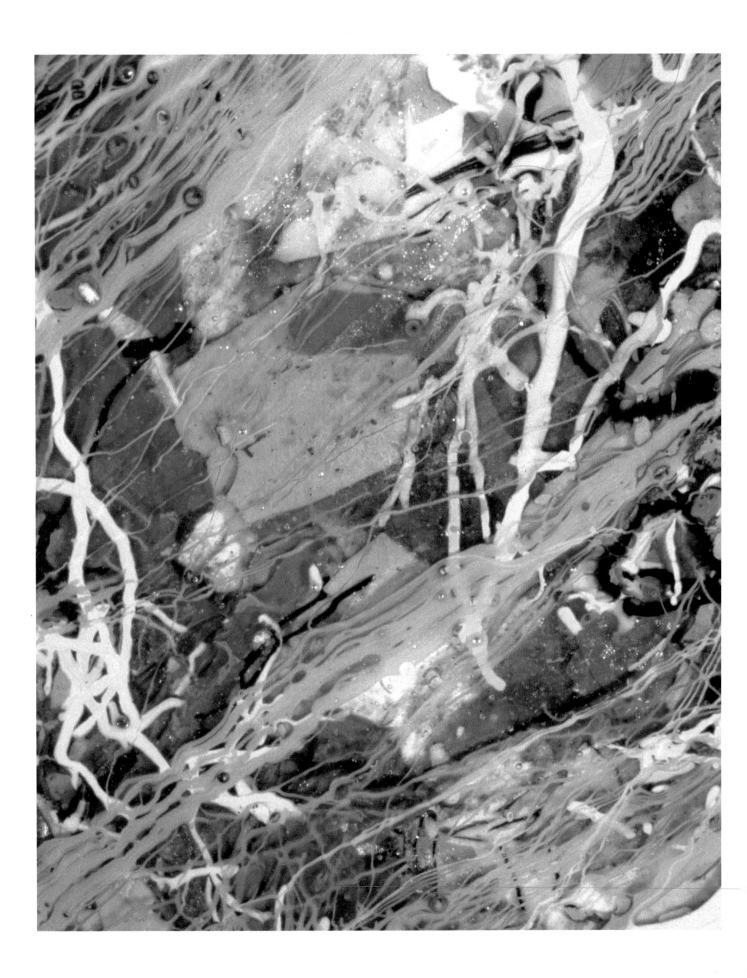

Crucible Casting

GLASS

The first realization that comes to my students when they observe crucible melting and the processes associated with it is, "That's how I can use all my scrap stained glass!" All kinds of glass can be melted in a crucible. In many instances, different glasses can be mixed, even though they have different coefficients of expansion. For many procedures, sorting by color and understanding a little about the oxides that make the colors is more important than knowing who manufactured the glass.

This is not true if the pieces are going to be fused to a specific glass. In that case, glass must be sorted by manufacturer. When casting singular pieces, pressing jewels, rolling sheets or making pieces that are to be foiled into panels, glass should be sorted by color. Bullseye (90 coefficient of expansion), Kokomo (93-94 coefficient of expansion), Spectrum (94-96 coefficient of expansion), M.G.R. (95-97 coefficient of expansion) and colored glass from all other manufacturers should melt well together if they are broken up into nickel and quarter-sized pieces. Melting these glasses together will give an average of all coefficients of expansion and most of the colors can be blended to form new colors.

Glasses containing selenium (red and orange) and cadmium (bright yellow) can be mixed to achieve various shades, but they should not be mixed with most other colors, since they will turn brown or black. Small amounts of any glass containing lead, when mixed with a selenium red or orange, will cause the selenium in the glass to "drop out", creating a dark, muddy color. Therefore, gold pink or gold ruby, which contain lead, should not be added to selenium red glasses.

Turquoise or aqua, those colors made with the metal copper, can also "drop out" of solution when mixed with cadmium glasses or selenium glasses. But copper blues can be mixed with cobalt blues. All other colors can be mixed and all colors can be lightened with clear. Cathedrals can be mixed with opals and white opal can be mixed with any color.

The concept is to crush the different glass cullets (scrap) and melt them into one homogeneous mixture and color. Don't expect mixed colors to come out marbled or streaky. When mixing different manufacturer's glasses, streaks of different colors may mean the glasses were not melted at a high enough temperature, and have not become one glass. Striations in the melted glass could indicate that glasses of differing coefficients of expansion have not blended, creating a new coefficient, and the situation would result in unannealable stress and subsequent breakage.

Different glasses have different working properties. This is due to formula components included to affect color, as well as formula differences made to affect forming characteristics. Machine made glasses are quite different than hand cast glasses. Bottle glass melts at higher temperatures than colored sheet glass. It also has a shorter working time. It is formulated specifically to have properties that work in bottle making machines. These working properties have little to do with coefficient of expansion.

Opposite page: Closeup, "Architectural Tile #4", Boyce Lundstrom.

Have bricks at hand as a resting place for the hot kiln lid.

Getting a good grip on a crucible holding twenty pounds of glass, using homemade two-man tongs.

The working properties of glass are very important to the glass artist, but can only be affected in a cullet melt by making raw material additions or by mixing glasses with different working characteristics that are understood. The temperature of the glass during melting, casting, pressing or blowing has the most effect on the finished product and is the easiest to control.

Artists who melt glass talk about "short" glass, "long" glass, quick "fining", and "sweet" glasses. These terms are most often used by glass

Crucible Casting

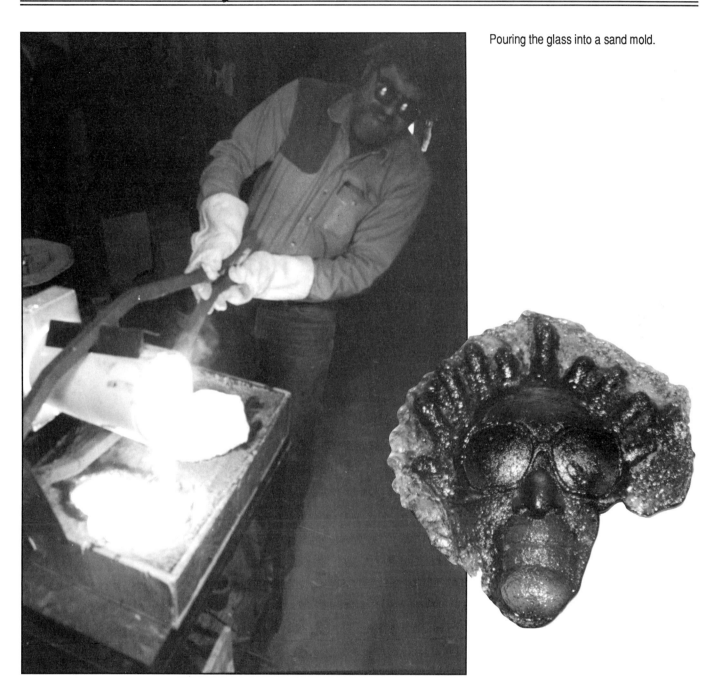

Pouring the glass into a sand mold.

blowers who are formulating glasses from batch. Since we are considering remelting various kinds of glass cullet, it is best to concern ourselves with the effect of temperature on the melt.

The following chart is very general. It is meant as a starting place for those who want to melt combinations of various sheet glass cullet.

2300°-2600°F
    "Fining" temperature for mixed glass "batch".
2200°-2300°F
    Melting temperature for mixed glass cullet.
2000°-2100°
    Casting temperature or pouring temperature from crucible.
1900°-2000°F
    Press molding temperature.
700°-1900°F
    Gathering temperature for blowing.
960°F
    Anneal soak temperature for mixed glass cullet.
1050°-850°F Annealing range for mixed glass cullet.

When melting combinations of sheet glass cullet used for casting, you will find most combinations will contain more color or more opalescence than is desired for the cast object. Clear glass used for thinning the color and density will quickly disappear from your scrap box. It is not necessary to break up sheets of glass to lighten the colors. Just add clear bottle glass cullet.

Clear bottle glass has varying coefficients of expansion between 88 and 92, with an average of about 90. It melts at a slightly higher temperature than most sheet glass and is stiffer (less runny) at an equal temperature. If crushed or broken to the size of nickels and dimes and mixed into other scrap, it is a great addition. When the bottle glass is more than fifty percent of the total mix, add 10% borax for all the bottle cullet added.

MEDIUM BLUE FOR CASTING
8 lbs. mixed blue cullet
12 lbs. crushed clear bottle glass
1.2 lbs borax (5 mol.)

Place 5 to 6 lbs. of glass at a time in a 5 gallon plastic bucket. Crush the glass with a heavy piece of metal welded to a long pipe handle. Place crushed glass in a second bucket. When all glass is crushed, add a splash of water and roll the bucket to dampen all glass. Add borax and roll again. *The better the mix, the better the melt.* This general formula works for all glasses and all colors. The borax and water facilitates melting and fining.

Melting time depends on the amount of glass and the temperature of the crucible furnace. Twenty pounds of glass should melt in two to three hours if the furnace is already hot. Mixing helps all glasses. At 2200 to 2300°F, a stainless steel punty rod can be pushed very easily to the bottom of the crucible and rotated in a stirring motion.

MELTING FUSIBLE BULLSEYE

Melting glasses in a crucible without changing their coefficient of expansion is important to those who intend to fuse the resulting shapes with similar glasses. For example, making design bars or trailings that can be fused to Bullseye fusible glass is often the goal of my students. Sometimes they want to make marbled glass or sheets with distinctly different colors. These processes are fairly simple to do, if you keep all your scrap separated in the studio, and if you follow a few general rules.
1. Crush only tested compatible glass.
2. Don't heat the glass over 2100°F.
3. Use the glass as soon as it is workable.

Glasses that sit for any extended time at elevated temperatures will volatilize. As the fluxes boil off, the coefficient of expansion changes (lowers).

To make marbled glass, one color or a mixture of compatible colors are placed in a crucible and melted to 2000°F. (A mixture of colors will blend at 2000°F and become one color.) A crucible filled with cullet will be one-half to

Crucible Casting

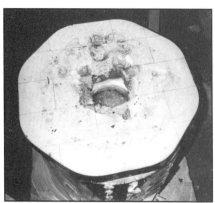

Joan makes a free form trail as the hot glass runs off of a gathering iron.

2000°F and become one color.) A crucible filled with cullet will be one-half to two-thirds full when melted—just right for adding the marble color. Strips of compatible glass equal in length to the depth of the crucible are prepared. The furnace is turned off, a bundle of the strips held in one gloved hand and pushed into the melted glass until it touches the bottom of the crucible. This can be done more than once depending on how marbled you want the glass. The furnace is turned back on and the glass brought up to casting temperature. As soon as the glass is ready (5 to 10 minutes), it can be poured. This is particularly nice using clear as the initial crucible melt.

Basic tools necessary for sheet casting.

Making sheets of glass can be very exciting, especially if some egg-shell-thin glass, prepared frit, and "streamers" (thick and thin odd-shaped stringer) are made beforehand to include in the glass sheets. The tools necessary are a metal table (cast iron works best because it does not warp when it gets unevenly hot) and a heavy metal roller. A pair of diamond shears or tin snips to cut the last tail of glass from the crucible and a large wooden paddle to transport the finished sheet to the annealing oven are necessary.

Before starting the sheet casting process, an annealing oven with enough kiln shelves to accommodate all the sheets that will be cast in one session should be set up. A front loading, side fired kiln works best. If using a top loading kiln, sheets can be stacked using pre-fired fiber paper between sheets. If sheets are stacked, they must be annealed for a long period of time because of the thickness of the sheet pile. Four sheets, 1/8" to 1/4" thick, weighing 4 to 5 lbs. each, with fiber paper between should be annealed at 930°F for three hours., then fired (ramped) down to 600°F over eight hours before the kiln is turned off. Zircar fiber shelves can be stacked on kiln posts between sheets when annealing in a top loading kiln, even though the kiln is hot, because the fiber shelves do not thermal shock. Sheets are loaded into a kiln holding at 1000°F.

Stringers, streamers, confetti and other thin design elements are laid out on a hot metal table. The table can be heated with a torch or by letting the first sheet of glass sit on the metal table until it transfers all its heat. (Do this only if you have lots of glass.) A hot table will afford some control of the thickness of the finished sheet. The glass is poured in front of the roller, and spread to the sides using a metal rod. The heavy metal roller is rolled over the hot glass one time only, pressing both sides of the roller down toward the table as the glass is rolled. The sheet of glass can then be grabbed with pliers and pulled onto a damp piece of plywood or thin metal sheet, and transferred quickly to the kiln.

As with all glass techniques, sheet casting requires practice. Two or three sheet casting sessions should be adequate to gain control of the process.

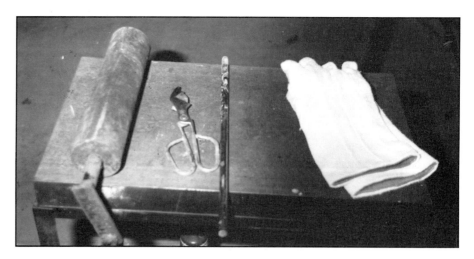

Crucible Casting

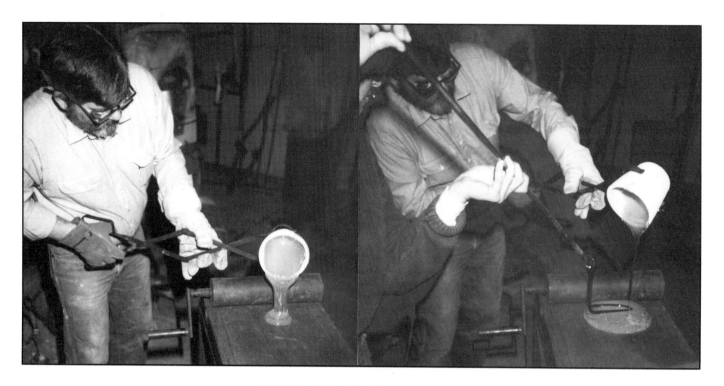

Glass at 2300°F is poured in front of the roller as contrasting color is trailed.

Mike experiences making his first sheet of glass.

Crucible Casting

COLORING FLOAT GLASS

Float glass (window glass) scrap is very inexpensive or often free. It is an excellent fusing glass, but the color pallet is very limited. It is possible, once you have a crucible furnace, to color float glass, maintaining a similar coefficient of expansion as the original float scrap. Float glass has a coefficient of expansion from 85 to 87 and Bullseye glass is 90. By using Bullseye scrap as the coloring agent, and melting at a high temperature to burn off some of the fluxes, the resulting colored float glass will be fusible to float sheet.

Mix 15 to 20 percent Bullseye cullet with crushed float glass. Melt at 2300°F for 2 to 3 hours after a full melt has been achieved. Stir occasionally with a stainless steel punty rod. Pour small sheets, make stringers, rods, frit, or free form lines to create a colored glass pallet that will fit float.

Additions of metal coloring oxides to float glass can also be made without changing the coefficient of expansion, if they are added in small amounts and melted correctly. The following list of oxide proportions and the colors they produce is only a starting point. By no means is it a complete list of oxides and combinations of oxides that can be used to produce color in float glass.

.2%-.6%	Nickel oxide	violet, smoky gray
.01%-.6%	Cobalt carbonate	blue, very dark blue
.5%-1%	Manganese oxide	violet
.2%-1%	Copper carbonate	greenish blue
.2%-.8%	Potassium dichromate	light green to yellow green
.5%-1.5%	Ferrous oxide	green to yellow green
.2%-.5%	Silver Nitrate	yellow to yellow green

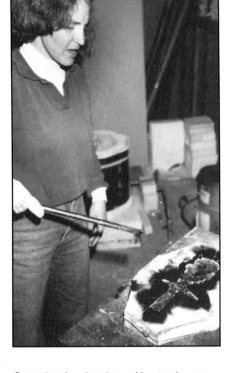

Super heating the glass with a torch over any trapped air bubbles will allow them to rise and pop.

Some other colors are produced by a combination of oxides and/or require a modifier, such as lead, in the glass batch to produce a distinct color. For example, it is not possible to get pink or red color from gold chloride, or yellow from cadmium by adding them to crushed float cullet.

When introducing metal oxides to crushed float glass, it is very important to crush the glass as fine as possible. Dampen the glass with water, then add the powdered oxides and mix very well. An addition of up to 10% borax (5 mol.) may be added to the cullet before melting. This will give a quicker, more even melt, but the mix will have to be melted longer to burn off some of the sodium introduced by the borax.

Building a crucible furnace and melting scrap sheet glass, or recycling bottles and float glass are logical steps to the advanced fusing artist who wants more control of his medium. The melting of glass cullet doesn't require any glass chemistry or great financial outlay, and is a way to recycle much of the glass that can't be used in other ways.

Crucible melting is not complicated. Anyone who can get together the materials can do it. And melting glass in a crucible, then pouring it out or manipulating it on the end of a punty rod is as easy as it is exciting.

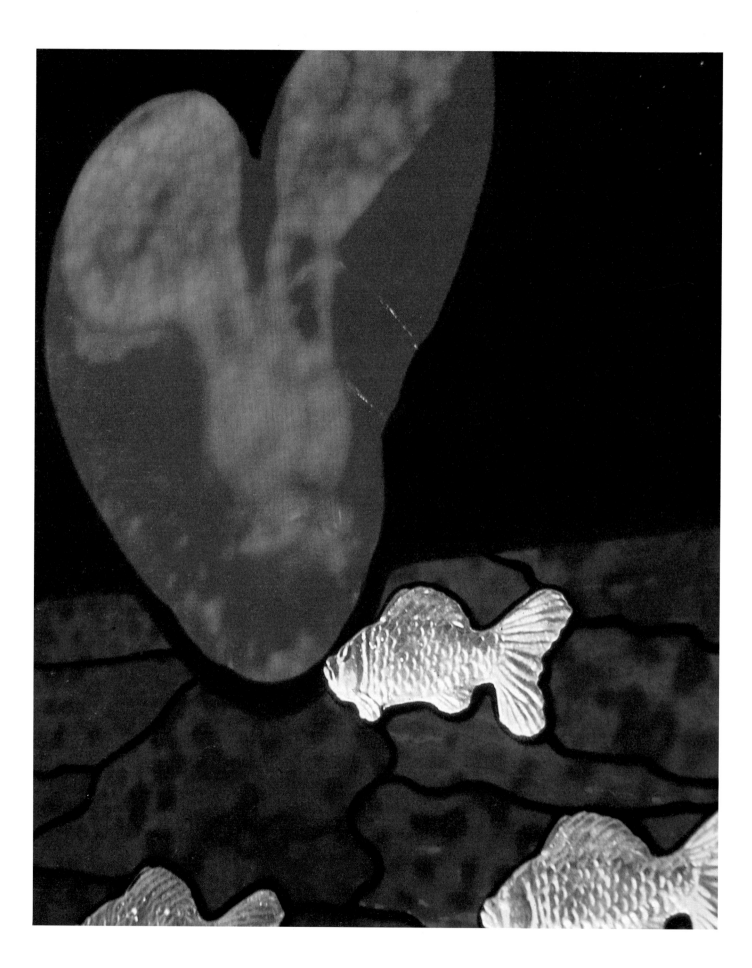

CHAPTER FOUR
Castable Mold Materials

CASTABLE MOLD MATERIALS

There are as many different formulas for castable mold materials as there are articles written on the subject. The problem with formulas is that reading the formula doesn't teach you the methods for using the material. You can have the best formula and materials but if you don't understand how to mix it, how to dry it, how to apply it, the formula doesn't really do you a lot of good. The total recipe and working knowledge are important. Success with castable mold materials, for use with glass, takes practical knowledge gained from experience.

The following chart organizes the most common materials used in castables by the function they serve:

BINDERS/CEMENTS	REFRACTORIES	MODIFIERS
Gypsum plaster	Silica	Vermiculite
Hydroperm	Diatomite	Pearlite
Hydrocal cement	Hydrated alumina	Grog
Portland cement	Zirconia	Sawdust
Calcium alumina cement (fondu)	Olivine sand	Air (soap)
Colloidal alumina		Alumina fiber
Colloidal silica		Luto
Clay (fire clays)		Kaolin clay

Many of the constituents in a castable mold mix overlap into more than one area of function. Fire clays can act as a binder or as a main body refractory. Some materials are available in many forms. For instance, plaster (calcium sulfate) is a general category. United States Gypsum Company manufactures many different plaster compounds under different trade names, each having unique working properties for specific industrial uses. Trade names of U.S.G. products will be used in all formulas in this book where they are called for. If a formula calls for "plaster", it refers to a No. 1 pottery plaster, which is the standard of the industry. Plaster is a binder, but also functions as a low-duty refractory.

BINDERS

The materials in a castable that hold the particles of refractory (high temperature material) together are called binding agents or cements. Different binders break down at different temperatures. Portland cement breaks down at 1600°-1900°F. Cement fondu or calcium alumina cement has holding strength up 2800°F. Most plaster compounds lose strength at 1300°-1500°F. Colloidal alumina and colloidal silica have binding strength from 800°-2300°F. All cements have varying degrees of strength at high temperatures, depending on the particular mix of refractories with which they are associated. Binders gain their holding strength three different ways:
1. Hydraulic bond (packing).
2. Molecular attraction (van der Wall).
3. Glassification or ceramic bond.

Opposite page: Closeup, "Coming into the Sea", cast fish, Boyce Lundstrom.

GYPSUM PLASTERS

USG #1 POTTERY PLASTER is finely ground gypsum mineral (calcium sulfate), which has been calcined to produce a uniform product. By changing the dehydration (calcination) and particle size, plasters are made to perform differently. United States Gypsum No. 1 Pottery Plaster is the standard of the industry. It is smooth flowing and strong. Other plasters may be substituted for No. 1 pottery plaster with slightly varying results. Quicker setting times, different shrinkage, the amount of water needed for proper crystal growth, and toughness can be affected by substituting a different plaster.

Some of the U.S.G. products that are commonly found in art stores, industrial supply outlets, and building supply houses, and can be used as an alternative to No. 1 pottery plaster, are as follows:

MOLDING PLASTER (referred to on the West Coast as casting plaster) is a good utility plaster where expansion control, hardness, and strength are not of major importance.

WHITE PLASTER is similar in working properties to molding plaster, except that it contains a surface hardening agent which minimizes paint absorption.

NO. 1 CASTING PLASTER is similar to white plaster, except that when it is mixed with less water it has greater strength, chip resistance, and further minimizes paint absorption.

POTTERY PLASTER is a general purpose plaster, commonly used for slip casting applications in the ceramic industry.

Any of the aforementioned plasters can be substituted for No. 1 pottery plaster wherever a formula in this book calls for it, although different amounts of water may be needed to achieve the same amount of dry strength. Since the dry cured strength of plaster differs so radically from the fired strength, and since, in all of the formulas that I have tested, refractory fillers and/or modifiers have been added, no definite negative or positive effects could be attributed to the kind of plaster, as long as it was a general-use gypsum plaster. Hydrocal and Hydroperm do affect the fired result noticeably, and should not be substituted directly for plaster.

HYDROCAL (WHITE) is a white gypsum cement that can be carved or added to. Setting expansion is about twice that of molding, casting, or pottery plaster. Hydrocal expands uniformly in all directions and has the highest setting expansion of any known gypsum cement. Expansion can be controlled by the quantity of water used in the mix. Expansion plaster allows for easy removal of models. For even easier model removal, rigid patterns can be covered with one or two coats of lacquer, followed by a coating of plaster-parting compound, such as stearic acid and kerosene. This will not alter detail or soften the face of the plaster cast. Total expansion for best model release takes approximately two to three hours after initial set.

Castable Mold Materials

HYDROPERM GYPSUM CEMENT is used to produce permeable plaster molds for casting non-ferrous metals. It contains a foaming agent that can be controlled by the amount and kind of mixing. Hydroperm is very permeable due to small, uniform cells (bubbles) that are interconnected. When the mold has dried, they offer channels through which steam or other mold gases may escape during the casting process.

TYPICAL PHYSICAL CHARACTERISTICS OF VARIOUS PLASTERS

Product	Parts of water by wt. per 100	Set time/mins	Dry density lb/sq ft
No. 1 Pottery Plaster	70	27-37	69.0
Molding plaster	70	27-37	69.0
White art plaster	70	27-37	69.0
No. 1 casting plaster	65	27-37	72.5
Pottery plaster	74	27-37	66.0
Hydrocal (white)	45	25-35	90.0
Hydroperm	100	12-19	

PORTLAND CEMENT is the most common cement found at building supply stores. It is commonly used in conjunction with plasters to affect the set time or increase the dry strength of the plaster, but is not commonly used as the main binder in any high temperature (over 1200°F) mold. Portland cement becomes active next to a hot glass surface and readily sticks when used in bonding proportions. Portland cement is often used in insulation refractories that back up other more stable materials and are away from glass or flame contact.

CALCIUM ALUMINA CEMENT (FONDU) is the most common bonding agent used in air setting castable refractories. It maintains its bonding strength up to 2600°F. An addition of up to 15% calcium alumina cement may be added to plaster to increase the mold mix's high-temperature strength. Additions of over 4% require the addition of small amounts of potassium sulfate in order to induce setting in a reasonable time. Calcium alumina cement is the binder in Kastolite 20 and Kastolite 25. These refractory, light-weight castables are generally used to back up hard brick on furnaces, and some craftsmen use these commercially prepared castables for slump molds. The strength of the calcium aluminate bond is so strong that glass cannot be removed from molds, without breaking the glass, if there are any undercuts.

COLLOIDAL ALUMINA is a suspension of finely divided alumina particles in a liquid medium. It is often used as a bonding agent for fiber products. It can also be used as a mold separator, ie. painted or sprayed on the surface of a mold to facilitate glass separation from the mold. It can also be used to bond olivine sand that has been packed tightly.

Original clay models in a cardboard box ready for plaster mold material.

COLLOIDAL SILICA is similar to colloidal alumina but does not have the same glass release properties. Both colloidal silica and colloidal alumina will air set, but achieve their greatest strength when fired over 800°F.

CLAY, fire clay, china clay, ball clay, and E.P.K. (Edger Plastic Kaolin) are used as refractory fillers in many castable mixes but not as binders. Most clays must be fired to 1600°F or higher to achieve a good ceramic bond (glassification). At this temperature they become very hard and tough. Any areas with undercuts in a clay mold will not permit glass to release easily. All moist, prepared clay compositions that have been bisque fired (1700°-1800°F) can be used as slump molds or as open-faced containment molds, but since they do not shrink or give when the glass contracts, only modest surface decoration can be used. E.P.K. is generally added to plaster mixes to help release the glass from the more active plaster surface. All clays are alumina silicate in composition and their function in plaster compounds is to create a variety of particle size to facilitate packing. Fire clays are coarse: 120 to 250 mesh; ball clay and E.P.K. are very fine: 250 to 400 mesh.

Castable Mold Materials

REFRACTORIES

Refractories are materials that will withstand high temperatures without deforming or changing chemically. In glass casting mold mixes, these materials form a major portion of the castable mold mix. Refractories have varying ability to resist the corrosive action of hot glass. In other words, some materials stick to glass more than others. This is due to their chemical formula, particle size, and the type of binder with which they are associated.

A variety of particle size of refractory material is very important to a mold mix that is cemented by hydraulic bond, such as plaster. Plaster decomposes as it reaches high temperatures. Its strength all but disappears at 1400°-1550°F, for many glass processes. As the plaster reaches 1200°F, it starts to shrink radically. This shrinking is what causes cracks in the mold. In the case of frit casting, the glass frit is becoming more consolidated at these elevated temperatures and the mold material (plaster) is shrinking. Refractories are inert; they do not shrink or change form. Therefore, if a plaster mold mix has 50% silica, it should shrink 50% less than if it was totally plaster. This works up to a point. After 50% silica or other refractory has been added, there is not enough bonding action to hold the refractories in place. By adding refractories of varying particle size, it is possible to pack the particles so closely that they have a certain amount of physical integrity. This packing of various particle sizes adds tensile strength by decreasing shrinking and, therefore, cracking.

SILICA may be added to mold mixes as flour (200 mesh and finer) or as sand. Silica sand can be obtained in various mesh sizes including 120 mesh, 80 mesh, 60 mesh and larger. Silica sand larger than 80 mesh may settle while the mold material is vibrated (a process used to release bubbles from the model surface.) Silica becomes active at elevated temperatures and sticks to the glass surface more than alumina or zirconia. Silica also goes through quartz inversion at approximately 1100°F and either expands or shrinks, depending on whether the mold is being heated or cooled. This can cause thermal shock cracking if the rise or fall in temperature is too sudden.

DIATOMATIOUS EARTH or diatomite is an exoskeleton of a tiny water plant. It consists mainly of silica and its crystalline structure is very open. It is lighter than silica flour and does not pack readily. It is an excellent additive to plaster mold composition in place of silica.

ALUMINA (HYDRATED). Alumina occurs in several crystal forms. These forms have the same chemical formula but differ in the way they respond to heat. Hydrated alumina with a particle size from 350 to 325 mesh is commonly used as a 25-50% additive to kaolin clay as a shelf primer. In larger mesh sizes, its thermal shock resistance make it an excellent choice as an addition to dense, cast open-faced molds intended for multiple-piece reproduction from one mold. Hydrated alumina and alumina silicate can be bonded with colloidal silica, resulting in very strong molds with low thermal expansion and excellent surface reproduction.

ZIRCONIA. At high temperatures, around 3180°F, zircon disassociates to form zirconia and silica. Zirconia goes through a phase change much like silica does, causing expansion, although this takes place at much higher temperatures than those to which glass molds are ever subjected. Zirconia is very stable at glass molding temperatures and does not react to the acid attack of glass. It is available in particle sizes from 30 mesh down to 200 mesh and finer. For this reason, it can be used when a variety of particle size is required for mechanical packing to increase fired strength.

OLIVINE SAND is made from naturally occurring minerals. Fosterite and fagalite (magnesium iron silicate) are mined and ground to sand-sized particles. It is most often used in metal casting sand mixtures. As a fine particle, is is a good addition to plaster-bonded mold mixes. In appropriate mixtures, it is often used as a direct glass mold. Foundry sands are often classified with an A.F.S. number, which refers to a mix of various mesh sizes. Therefore, olivine 120 does not refer to 120 mesh but a mixture of sand particles from 50 to 180 mesh. Olivine sand, unlike silica sand, does not go through quartz inversion. It has a low thermal expansion rate. Olivine sand can be bought through foundry suppliers.

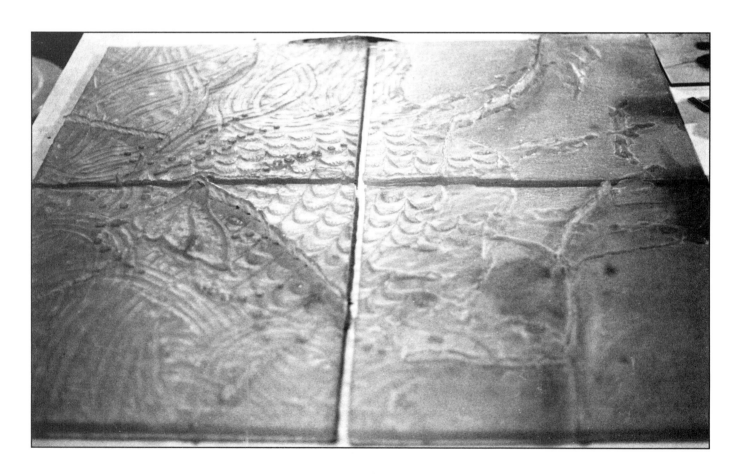

Mike Dupille's clay model made in four sections, each 20" x 20".

Castable Mold Materials

Is he having fun yet?

MODIFIERS

Various materials are added to castable glass mold formulas to enhance drying time, to increase porosity, or to absorb expansion or contraction of other mold materials. They do not add bonding strength; in fact, they often weaken the mold structure. Whereas they may be refractory, their addition to the recipe is usually in small amounts and not as a main body filler. The misunderstanding of modifiers and their overuse in many glass mold formulas probably causes more mold failures than any other reason.

The following discussion of modifiers is meant to add to your general knowledge. Some, but not all, of the modifiers discussed are used in mold recipes in this book.

VERMICULITE is a clay mineral, which looks like mica. It is expanded by heat, which puffs it to 10 to 15 times its original size, making it very light and airy. It is dark gray in color. It is used commercially as an insulation and as a soil conditioner. When used as a glass mold modifier, it should be rubbed and pushed through a window screen (approximately 10 mesh).

PERLITE is a naturally occurring mineral, expanded with heat very much in the same manner as vermiculite. It is white or light gray in color and used commercially as an insulator, soil conditioner, and a light abrasive. It is finer than vermiculite, but still should be screened to 10 mesh or finer before use in glass mold mixes.

A mold carved from dense fiber board, treated with colloidal alumina.

The cast object is released without destroying the mold.

Pouring large molds can involve a helper and good timing.

GROG is ground pre-fired clay. It is available in various particle size from 10 mesh to 200 mesh. Ione grain sand is a naturally occurring mineral often substituted for grog and at times referred to as grog, but it is not a good substitute. Grog lowers expansion and contraction rates.

Castable Mold Materials

SAWDUST, wood fiber or raw paper have sometimes been used as an addition to mold mixes to create a more porous, open mold. The wood fiber burns out of the mold mix during initial firing, creating air vents. These materials are difficult to control in gypsum cement formulas in that they absorb water, therefore affecting the crystal growth and strength of the gypsum cement bond.

AIR can be entrapped into a mold material by the the addition of a foaming agent. Soap or detergent can be added to mold mixtures and, when properly agitated during mixing, creates a fine interlocking bubble network that helps porosity. Hydroperm contains a foaming agent, and can be used as a component of various mold mixtures.

Mike Malone holds the kiln lid open while glass is added to the hot mold.

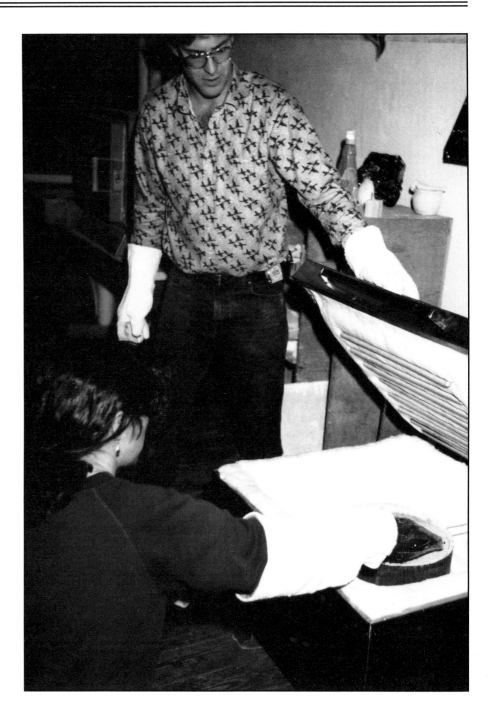

ALUMINOSILICATE FIBERS such as Fiberfrax, Kaowool, and Cerafiber, made by Carborundum, Babcock and Wilcox, and Johns-Manville respectively, can be obtained in blanket form, chopped fiber, or ground fiber. These fibers are very refractory and can be bonded with colloidal silica or colloidal alumina to form a glass mold by themselves. They can also be used as modifiers in castable glass molding mixtures. They add tensile strength, resistance to thermal shock, and porosity to gypsum molds as well as to clay molds.

Castable Mold Materials

LUTO, or ludo, is a name given to crushed, previously fired mold material. It is used as an inert filler in much the same way as vermiculite or perlite. When I was in El Salvador, the soil was referred to as ludo, and it was used for bricks, for pots, and to cover houses made of thatched sticks. I think at times it was used as a seasoning in rice stew. I much prefer the idea of reusing a small amount of the past mold in each new mold, rather than using vermiculite. The material should be crushed and sieved through a 10 mesh screen. When used as an addition to a mold mix, after liquidation and just before pouring, ludo makes the mold less likely to crack due to fast kiln drying. It also makes it taste good.

KAOLIN CLAY is very, very fine alumina silicate. An addition of 3% to 5% to gypsum molds help separate the glass from the plaster.

ANALYZING CASTABLE RECIPES

There are as many different reasons for castable formulas as there are formulas. Someone publishes a formula because it worked for him. If you understand what *function* the materials have in any castable mold mix, you can modify the recipe to work with your particular process. It is often necessary to alter castable formulas, because there are so many variables in individual glass work. The type of glass being molded is one variable: soda-lime, soda-lime borosilicate, lead, or barium-based glasses all respond differently to the mold material. Some castables pick up fine detail and some don't. Look closely at failures as well as the successes of molds, and if a mold mix is not satisfactory, modify it. Successful glass molding is simple, but it is not easy. Procedure, timing and technique are learned by practice.

Multiple sun faces taken from one ceramic mold.

Mike Dupille, "World in the Middle of a Flower", slump cast over carved plaster.

CHAPTER FIVE
Mold Formulas

CASTABLE MOLD FORMULAS AND PROCEDURES

General procedures for mixing plaster are the most important background knowledge any moldmaker should understand *thoroughly*, before starting any mold project dealing with plaster-bonded mold mixes. Much of the following material was taken from the United States Gypsum Company copyrighted bulletins No. IG-503, IG-538, IG-539, IG-540, and IG-502, and is used here with their permission. I have tried to sort out technically pertinent information that relates to the studio mold maker and, specifically, information that I have found helpful in teaching mold making to glass artists.

The ideal plaster mix is one in which the plaster particles are completely dispersed in the water to produce a uniform, homogenous slurry. Care must be taken to *control variables* such as batch size, mixer design, mixing time, water purity and temperature, and the use of additives.

Water purity and temperature affect the setting time and surface quality of the finished plaster mold. If the water is drinkable, it is suitable for mixing plaster slurries. Plaster has maximum solubility at approximately 100°F. It also has the shortest setting time at 100°F, which can cause problems if the mold maker is not in control of this variable. Molds made in the winter in a cold studio and in the hot summer months, using the same procedure, will definitely produce different results. Most studios have hot and cold water, so mix your water to a *known* temperature every time. I suggest 70°-75°F for *all* mold mixes, except Hydroperm. The temperature of the dry plaster can also be a variable.

The water-to-plaster ratio will affect the strength of the plaster and, therefore, the mold performance. No. 1 Pottery Plaster requires 70 parts water to 100 parts plaster for best results, yet R&R 965 investment calls for 28 parts water for 100 parts plaster. This dynamic difference should point out the need for weighing of all materials. The addition of foreign particles, ie. vermiculite, sawdust or silica affects the need to change suggested water ratios. This is not as difficult as it sounds. If the added material absorbs water, increase the water; if the added material is inert, leave the water the same.

Soaking, or slaking, the plaster after it is introduced into the water is very important. Soaking allows each plaster particle to be completely saturated with water so that it is easier to disburse. Plaster mixes become completely wetted after two to four minutes. Short cuts in soaking will negatively influence the effectiveness of the mixing and, consequently, the quality of the finished glass mold.

Mixing the plaster slurry is the most important step in producing a quality mold. Variations in mixing procedure will have a greater effect on the finished product than will any other phase of the process. There is a direct relationship between energy input during mixing and strength development of the cast. As mixing time increases, strength of the mold increases. However, long mixing time will adversely affect porosity of the mold. In molds for glass, porosity is very important. The size of the mold batch and the type of mixer used affects the energy input and, therefore, the mixing time. The general rule of thumb is: when mixing batches using five to seven pounds of water, slake the plaster, mix two to three minutes and stir with a Jiffy mixer for two to five minutes.

Opposite page: "Moon Deer Renewal", 41" x 18" x 7 1/2", frit cast glass, metal and granite, Ruth Brockmann.

JIFFY MIXER

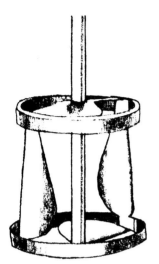

MODEL ES
21" SHAFT

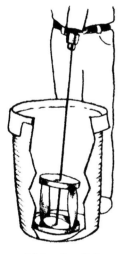

MODEL PS
40" SHAFT

The mixing container should be clean and sturdy. The most common mold failures in my classes are ones caused by use of unclean equipment in the mixing. A bucket and mixer left inadequately cleaned at the completion of a mix will allow the plaster remaining on the mixing blades and the side of the bucket to accelerate the set of the next batch of mold mix. This can cause the plaster mold mix to set in one half the expected time, and even prevent a complete reproduction of the model.

It may be helpful to summarize general directions for mixing most plaster mold mixes as follows:

1. Start with a very clean bucket and mixer.
2. Weigh water and plaster mold mix before combining them.
3. Use clean tap water at approximately 70°F (use a thermometer).
4. Sift or broadcast the plaster mix into water slowly and evenly. DO NOT drop handfuls of plaster directly into water.
5. Allow the mixture to soak two to three minutes before stirring.
6. Mix with Jiffy mixer for two to five minutes.
7. After pouring the mold, immediately clean equipment thoroughly.

Here are a few tips on cleaning equipment. In our studio we use four six gallon plastic food buckets obtained from restaurants. One bucket always has clean water for mold mixes and that bucket is not used for *any* other material. A second bucket has water for an initial cleaning of equipment. A third is used to hold waste water, and a fourth is used for pouring unused plaster mix into. It is lined with a plastic trash bag. After mixing, the Jiffy mixer is run in the cleaning water before it is set down. After pouring the mold, the excess plaster mix is immediately poured into the plastic lined bucket and cleaning water is poured into the mixing bucket. After it is swirled around, this water is poured into a third bucket. The mixing bucket and the Jiffy mixer are wiped with a rag.

Within a few hours the plaster in the waste water bucket will settle and the water on top can be poured into the cleaning water bucket. The slurry can be poured into the trash bag. Why is this important? Because even small amounts of plaster can plug drains, even a sump system. Also any amount of plaster in the mixing water will accelerate the plaster mold mix. Plaster water is hard to get rid of without causing unsightly white areas on the ground, but don't pour it down the drain. Even small amounts of plaster will settle out and harden and can be thrown in the garbage. Dig a hole next to your studio to pour the water in, or pour it on the compost pile. Used plaster molds and plaster water are great soil conditioners, if your soil needs calcium.

MODEL	Shaft Length in.	mm	Container Size	Top Diameter in.
LM	10 1/2	267	pint	1 1/4
HS	10 1/2	267	1-2 gal.	2 3/8
HS	15	381	1-2 gal.	2 3/8
HS	30	762	1-2 gal.	2 3/8
ES	21	533	2-5 gal.	3 1/4
PS	21	533	5-10 gal.	4 1/2
PS	40	1016	10-50 gal.	4 1/2
HD	40	1016	100 gal.	9

SIZED FOR A RANGE OF USES...

Mold Formulas

The following mold recipes are suggested for open faced, one-part molds. They have a fine grain, may be carved before curing and pick up and maintain fine detail. *All mixtures are by weight unless otherwise stated.*

MOLD MIX 50/50
1 part No. 1 Pottery plaster
1 part Silica flour

Mix dry ingredients together well before adding them to the water. Use 50 parts water to 100 parts mix. Cure the mold using the slow cure method. When making large molds, add one cup of loosely packed chopped fiber to five pounds of water, mixing the fiber into the water before sifting in the dry ingedients. Fiber can inhibit carving detail.

MOLD MIX DIATOM 50/50
1 part #1 Pottery Plaster
1 part diatomatious earth
5% shelf primer or E.P.K.

Mix dry ingredients together before adding them to water. Use 60 parts water to 100 parts of dry mix. This mix is very easy to carve and has good release properties for lead glasses. Curing may be faster than for Mold Mix 50/50, when mold walls are 1 1/2 inches or less. This mold material has shown satisfactory results up to 1550°F.

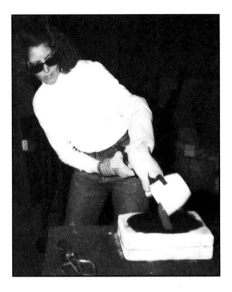

Wire wrapped around very deep molds adds strength as they dry.

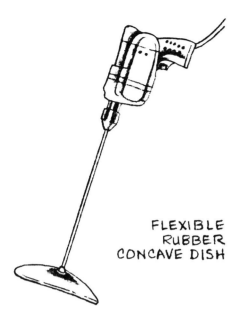

FLEXIBLE RUBBER CONCAVE DISH

MOLD MIX H 80/20
4 parts Hydroperm™
1 part diatomatious earth
5% E.P.K. Kaolin clay

    Mix dry ingredients together before adding to water. Use 80 parts water to 100 parts of dry mix. Hydroperm should be mixed with a disc mixer, which will beat air into the mold material, making it increase in volume by as much as 70%. Hydroperm contains a foaming agent and, when mixed properly, will produce cells approximately .01 inch in diameter. Straight Hydroperm mixed 1 to 1, by weight, with water will increase 100% by volume. Since Mold Mix H 80/20 has an addition of 20% diatomatious earth, expect only a 60 to 70% increase in volume.

    The following mixing instructions are for quantities that start with five to eight pounds of water, mixed in a five-gallon plastic bucket. Smaller amounts do not work as well without changing the size of the mixing container and the disc.

    1. Start with 100°F water.

    2. Add H 80/20 by strewing the dry material over the surface quickly, taking approximately 15 seconds.

    3. Let soak for 30 seconds, then mix with your hands until the slurry is free of lumps.

    4. Place the disc mixer approximately two inches from the bottom of the bucket and turn it on. The speed should be sufficient to create a vortex of air down to within an inch or two of the bottom of the bucket, when the mixer shaft is in a vertical position.

    5. When the desired volume increase is obtained (approximately one minute for five-pound water mixes), raise the rotating disc to just below the surface of the mix. If a variable speed drill is used for mixing, operate it at a very low speed. There will be many large bubbles in the slurry that need to be reduced. Raise and lower the disc while operating at slow rotation speed, for one minute.

    If you have read the instructions for Mold Mix H 80/20 from the beginning you may be thinking, "What a lot of special procedures!". It does take practice to develop the skill to make a suitable mix that contains air bubbles of proper size (.01 in. diameter). But it is worth the effort, especially if you like to work fast and you desire a mold material that can be removed from the mold in one hour. The model can also be removed at that time and the mold can be placed in a kiln or drying box immediately. Drying takes approximately three hours at 350°F, for a ten-pound mold. Additional time or higher temperatures are necessary for larger molds.

    Cool the mold rapidly if you like, and load it with glass frit. If you are using the mold as a pour mold for hot glass, dry the mold for an additional two hours at 600°F, and cool the mold slowly. Mold Mix H 80/20 is very permeable, yet generally gives excellent surface quality, and has the ability to pick up very fine detail. This mold material has been used successfully up to 1600°F with all types of glass.

    Apply a parting compound to all types of patterns to permit withdraws. U.S.G. Company Bulletin No. IG-539 has more information on the Hydroperm process.

Mold Formulas

MOLD MIX 522
5 parts Hydroperm
2 parts Hydrocal
2 parts silica flour
2 cups loosely packed chopped fiber to 7 pounds water

Mix all ingredients except the chopped fiber before adding 75 parts water to 100 parts of dry mix. Add chopped fiber to water and stir in thoroughly with a Jiffy mixer, before strewing plaster mix over the surface. Use a disc-type, high-energy mixer and mix as described for Mold Mix H 80/20. It is not necessary to use water warmer than 70°F. Expect a volume increase of 30 to 40%. Fiber content can be doubled in batches with more than eight pounds of water, but is only necessary for large, flat molds. This mold material should be cured and dried slowly. Mold Mix 522 can be used for all types of glass and all processes, including pate de verre, open-faced molds, and lost wax molds. When cured properly, it has performed well to 1625°F. Ten percent previously fired and screened mold material (luto) may be added to the slurry before pouring. This would allow shortening of the drying time.

MOLD MIX R & R 965
calcium sulfate (Gypsum)
cristobalite
crystalline silicate
graded refractory materials

Ransom and Randolph manufactures an investment castable made specifically for aluminum and copper-based alloys. It has been used very successfully for pate de verre molds, as well as for open-faced molds. Mixing and curing instructions are provided by Ransom and Randolph. In them, it is suggested that 28 parts of water be added to 100 parts of investment compound and the mixture mechanically mixed for two to three minutes. The mold should be vibrated to remove air bubbles, which may adhere to the pattern. This castable cures quickly. After two hours the mold may be placed in a burn-out oven. The mold should be held at 250-300°F for four hours, then brought up to 1200°F, at the rate of 150-200°F per hour, and held at this elevated temperature for five hours. The exact time and temperature will vary somewhat, depending on the size of the mold.

MAKING A HIGH ENERGY MIXER FROM A JIFFY™ MIXER

Several variables will affect the efficiency of a high energy mixer: the bucket size, the disc diameter, the mixer speed and the volume of the plaster slurry. The disc diameter should be approximately 1/2 to 2/3 of the diameter of the bottom of the mixing bucket. Adjustable mixer speed can affect the vortex that is created while mixing and, therefore, the amount of air that is incorporated into the mix.

We have found it practical to use a plastic three-pound-coffee-can lid adapted to a Jiffy mixer, powered by a 3/8" variable speed drill. To adapt it, drill a 3/8" hole in the center of the plastic lid and cut an H pattern with a knife to replicate the width of the Jiffy mixer blades. Push the lid over the Jiffy mixer to the bottom of the blades—the blades keep the lid from spinning. An extra pre-cut lid should be kept on hand, since eventual tearing of one in use is inevitable. A more durable mixer can be made with a soft rubber disc attached to an 18" shaft.

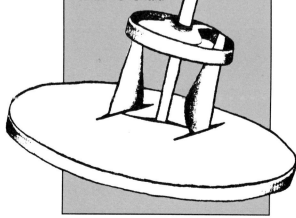

The slow heat up.

MOLD MIX HO 3/2
3 parts Hydrocal
2 parts olivine sand 180 (Washington)
5% E.P.K.

Mix ingredients together before adding them to water. Use 60 parts water to 100 parts of the dry mix. Let the slurry soak for two minutes before mechanically mixing with a Jiffy mixer for one to two minutes. This mix will set more rapidly than others in this chapter. After using a parting compound, paint the surface of the model with the mold mix, using a brush, before pouring the slurry down the side of the mold box. After two hours of curing, the mold may be dried using the fast method. This is an excellent material for slump molds and, if fired correctly (not over 1300°F), may be used repeatedly.

DURABLE PLASTER MIX
6 parts Hydrocal
2 parts silica flour
1 part grog (80 mesh)
1 part vermiculite per 3 pounds water

Mix all ingredients, except the vermiculite, before adding the 60 parts water to 100 parts dry mix. Soak for two minutes. Mix mechanically for 1 1/2 to 2 minutes. Add the vermiculite (screened through window screen) just before pouring. Vibrate mold box to release bubbles from the surface of the model. Use the slow curing process.

Mold Formulas

CURING AND DRYING PLASTER MOLDS

Gypsum plaster-based mold mixes need approximately two hours of setting time to develop proper crystal structure. After this initial set time, optimum physical properties are maintained by properly drying the mold. This involves a transfer of excess water from the cast to the surrounding air. Plaster requires about 18 parts water per 100 parts plaster, by weight, for complete hydration in the setting process. However, in order to obtain a usable slurry, greater amounts of water are added. After the plaster has been mixed, poured, and has set, any water above 18 parts is considered excess or "free" water and must be removed from the cast by drying.

Drying equipment can be designed to remove the excess water or the mold can be placed in the kiln to dry. The same drying action takes place whether the plaster cast dries in the workroom, outdoors, in the kiln, or in a mold dryer. Use of a forced hot-air dryer speeds and controls the drying procedure.

In our studio we built a dryer by constructing a box out of drywall (plaster board), with internal dimensions of 18" by 18" by 30". In one end a hole was cut large enough to insert a hot air gun designed as a paint stripper. The hot air gun has two settings: low and high. Five or six eight-to-ten pound molds can be placed in the dryer at a time, and the hot air gun set on low for 12 to 24 hours, depending on the amount of moisture in the molds. The temperature setting should then be increased to high and left at that setting until the air coming out the top and bottom vent holes will not fog a piece of glass or mirror when held in the exhausting air.

It is possible to reach 600°F in such a box, using a hot air gun set on high, after the moisture has been driven off. Drywall is nonflammable at this

Cross section of a plaster drywall box we use for drying molds.

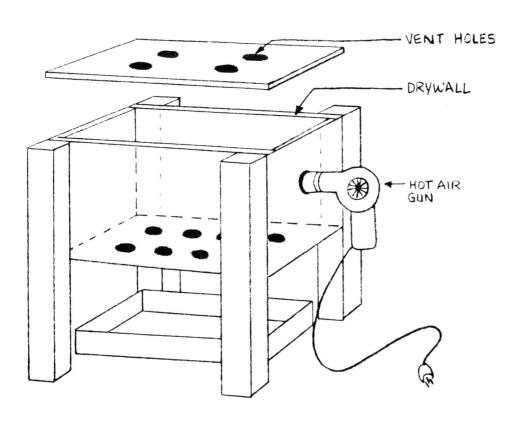

Take the time to dry your mold thoroughly

temperature, but standard safety precautions should be taken in placement of this chamber. For smaller boxes, a hair dryer may be used.

When a wet (new) mold is placed in a forced hot air dryer, rapid evaporation begins. This initial evaporation keeps the mold cooler than the air temperature in the dryer. Water from the interior of the mold moves to the surface to replace evaporating moisture. As evaporation continues, sufficient water does not move to the surface to keep it cool, and the surface temperature will rise, although the center of the mold is still moist. As the amount of evaporation is reduced, the mold's surface temperature approaches the air temperature in the dryer. Once the surface of the mold is up to air temperature, the rest of the free water in the mold evaporates slowly. When the center of the mold reaches the temperature of the surrounding air, the drying process is complete.

Molds with wax models inside them can be stilted approximately one inch over a hole in the bottom of the dryer. As the interior mold temperature approaches 200°F, the wax will melt and drop through the hole in the bottom, where it can be collected in a container for reuse. An air dryer does not get hot enough to totally remove the wax, so the mold should be brought up to 1200°F in a kiln to achieve total burn-out.

United States Gypsum Company Bulletin No. IG-502 suggests that temperatures over 120°F for No. 1 Pottery Plaster or Hydrocal will calcine the surface of the mold. This surface calcination will make the outside of the mold soft and powdery. Yet if temperatures lower than 120°F are used in the hot air dryer described above, drying time is increased five to ten times. Surface calcination does not negatively affect molds that are at least two inches thick around the cavity to be filled with glass. If problems do arise, in the form of soft or fragile molds, lower the temperature and lengthen the drying time.

Drying time varies with the thickness of the mold. The following table can be used as a guide for curing and drying time. These figures come from practical experience and are for molds of at least 2" wall thickness between the molded pattern and the outside wall, and of total weights of not more than fifteen pounds.

It is not uncommon to spend two weeks curing and drying a complicated mold that has had a lot of time and love invested in the contained art. Being safe is often less time consuming than finding it necessary to start over.

	MOLD CURING AND DRYING GUIDE		
	FAST	MEDIUM	SLOW
Curing time	2 hrs	12-20 hrs	48-72 hrs
Initial heat	250-300° F	200° F	120° F
Initial time	4 hrs	8-12 hrs	48 hrs
Final heat	300° F/hr rise	100-150° F/hr rise	50-80° F/hr
Final time	to casting temp or burnout	rise to casting temp or burnout	rise to casting temp

Mold Formulas

GENERAL MOLD NOTES-MOLD POURING AND THICKNESS

When pouring castables into a mold box, do not pour directly on the model. Pour the castable mix down the side of the box and let it flow over the model (pattern). This procedure should keep bubbles from forming on the pattern. Painting a thin slurry over the pattern with a fine camel hair brush before filling the mold box helps break the surface tension between the parting agent and the plaster mix. Vibrating the mold box by tapping the sides or using a vibrating sander held against the mold box helps bubbles rise from the pattern surface.

If the mold box is not held together with screws, but instead with tape or a clay dam, vibrating may cause the dam to break, letting the mold material run freely. If this happens to you once, you will change your building technique!

Molds of equal thickness in all directions dry evenly and, therefore, are least likely to crack. Two inches of castable around all parts of a pattern are

Heating the glass with a torch after laying in copper inclusions. More glass will be poured to cover them.

Plaster composition molds by Ruth Brockmann, some with copper cutouts set in place. They will be placed between layers of glass during the casting process.

Molds of equal thickness in all directions dry evenly and, therefore, are least likely to crack. Two inches of castable around all parts of a pattern are necessary for all plaster cemented castables. Because of irregularities of shape of patterns, molds should be cast large enough to contain the outermost part of the design by two inches, then cut to an irregular shape on a band saw after curing, but before drying. Exactness is not necessary, but removal of large pieces of unnecessary mold material certainly helps even drying. Don't forget the thickness of the mold from top to bottom, as well as from side to side. In the instance of casting a face or mask, where the mold has one protrusion (such as a nose), it is better to leave a flat surface for the mold to be supported on than to cut away castable to make the mold have even thickness.

Support molds greater than 10" by 10" on a sand base. Pour three or four cups of sand on a kiln shelf, then work the mold back and forth until it is uniformly supported.

Larger molds that may be used more than once, such as slump molds, should be ground flat and placed on their *own* kiln shelf. Always pick up the kiln shelf, not the mold. Slump molds that crack can be wired together and used many times after cracking, if they are handled correctly.

Level the mold table before you make your first mold. By doing this, many frustrating moments will be avoided. Level your kiln as well. Glass always flows level and 1/8" off level is very noticeable in a twelve-inch tile.

Most castable mold material is inexpensive. I feel you are better off mixing more than enough castable rather than too little. Throwing a little away won't break the budget, but it isn't fun to find you don't have enough for a complete cast, although most plaster bonded castables can be added to.

Mold Formulas

As a general rule of thumb, four pounds of water in a 60/100 ratio mix will give one cubic foot of mold material.

Variations in water-to-plaster ratio will affect the strength and performance of the mold. If molds crack or are fragile after curing and drying, check your notes on the amount of water used. Generally, less water will make a stronger mold. Chopped or milled fiber can be added to any mold mix to give it added tensile strength. Amounts of one-fourth cup to three cups per three pounds of water may be added to the water before mixing, or added and stirred into the mold mix just before pouring. Experimentation should be a part of your studio practice.

PARTING COMPOUNDS

When taking a mold off of plaster models, prepare the plaster surface by coating the model with two coats of lacquer followed by a coating of stearic acid, kerosene, or other parting agent. Most plasters expand upon curing, then shrink when fired. This expansion of mold mixes should help with removal of hard models if the surface is smooth, has been treated with a parting agent, and does not contain undercuts.

Suitable parting agents are described in United States Gypsum Bulletin IG-515. A parting agent should be insoluble in water, and fluid enough to form a thin film. Commonly used parting agents are listed below.
1. Stearine.
2. Polyvinyl alcohol.
3. Spray silicone.
4. Johnson's "Glo-Coat" and water.
5. WD-40.
6. Non-vegetable oil spray coatings such as Pam.
7. Soft soaps.
8. Light oils and emulsified oils.
9. Partlube 9-D (wax).

Parting agents should be applied sparingly to nonporous surfaces. Porous surfaces should be sealed. Heavy coats of any parting agent will penetrate the wet mold mix and will result in a poor mold surface and reduced sharpness of detail.

Stearine is a commonly used parting agent. It is prepared by melting down one pound of stearine and, after removal from heat, adding one pint kerosene. The material is brushed thoroughly over the pattern surface, but not so heavily as to show brush marks. If brush marks appear, heat the mold with a hair dryer and brush with more kerosene.

Generally, plasticine clays have enough oil in them that plaster does not stick, but the area around the plasticine clay needs to be covered with a separator, as does the mold box. Other soft clay materials should be coated with a light oil or wax spray. When soft clays or other pliable material is used for a model, undercuts in the model are not a problem, and the clay that does not readily release from undercut areas can be removed with picks or dental tools. White stoneware clay that does not get removed before the initial firing of the mold will shrink away from the plaster and become easier to remove. White clays (as opposed to red, iron clays) do not leave residue that may color the glass.

Removing the clay model from a plaster composition mold. The clay can often be used again after a little repair.

MODELS AND PATTERNS

Many materials work satisfactorily for making models of your intended glass object. Wet modeling clay or plasticine are usually a first choice. But unless you've had extensive experience working with clay, you may find it difficult to achieve more than a rough image. Clay does not hold detail easily.

True model making or pattern making is an art. Since many books are available on the subject, I will review the rudiments of the basic initial process that we use in our studio.

1. Make a model of clay, roughing in the general shape, and the bold detail.
2. Let the clay model dry to the leather hard stage, then reshape it, removing all undercuts and apply draft of 4 to 5%.
3. Place the model in the mold box, sticking it to the bottom board with clay slip.
4. Spray with "Pam" as a parting compound.
5. Make an original glass mold or make Hydrocal mold.

Mold Formulas

6. If an original glass mold is made, the process is complete as soon as the clay model is removed and the mold has cured and been kiln dried.
7. Remove the clay model, check Hydrocal mold for undercuts, apply wax parting agent to the cavity and cast a plaster pattern.
8. The plaster pattern can be carved with fine detail, cured, waxed, or varnished, then mounted to the base board of the mold box with hot glue and screws.
9. Many original glass molds may be taken from a properly prepared pattern.

Whereas the glass mold process takes about twice as long if taken to the completed pattern stage, the reasons to do so should be apparent. Molds crack during firing, glass application may not do the art justice or a piece may break during the cold working, finishing process. If the art is worth making, it is worth making a pattern that is semi-permanent. I understand *all* of the reasons for taking shortcuts, the foremost of which is usually a burning desire to see the original in glass!

There are modifications to this process that can make it more spontaneous without losing the original art work. A well drafted mold can be filled with wax or more clay (step 7). Both shrink and can be removed from the plaster easily. Many objects can be pulled from one well made mold, reworked, including the addition of undercuts in detail, then cast, making a glass mold. In this way the original work is not lost. Finished glass molds can be carved further before glass is added, changing the original basic shape. Carved surfaces should be cleaned with compressed air to prevent washing loose particles of plaster into the glass.

Patterns can be made of plastic, properly sealed Hydrocal, cement or wood. Wooden patterns must be properly waterproofed, preferably with lacquer. All rigid patterns must be designed with sufficient draft to permit release. File marks and rough surfaces may prevent extraction from rigid patterns. Most mold mixes suggested in the previous chapter are capable of holding fine detail and the glass surface will exactly replicate the mold surface. Flexible rubber patterns can be taken from molds where "back draft" is desired. Other flexible materials, such as cold molding compound, can also be used in "back draft" situations.

PATTERN EXTRACTION

Models and patterns become easier to remove as the plaster expands, upon curing. For most mold mixes two hours is sufficient. Molds can be thumped on the casting table to vibrate the model to help loosen it. Models made of rigid materials should be fastened to the bottom of the mold box. Remove the outside of the mold box and then, using a small wedge, tap it between the mold box bottom and the mold.

Compressed air applied at the right point is probably the best extraction method. Blow air between the mold and the bottom plate. Make a small hole through the plaster to the pattern with a thin wire and gently blow air through the hole. At times, vibrating the pattern is helpful. Use a hand held vibrating sander (without sandpaper), placing it on the pattern or the mold box. For flexible patterns, such as cold-formed rubber, blow air between the mold and the rubber side walls.

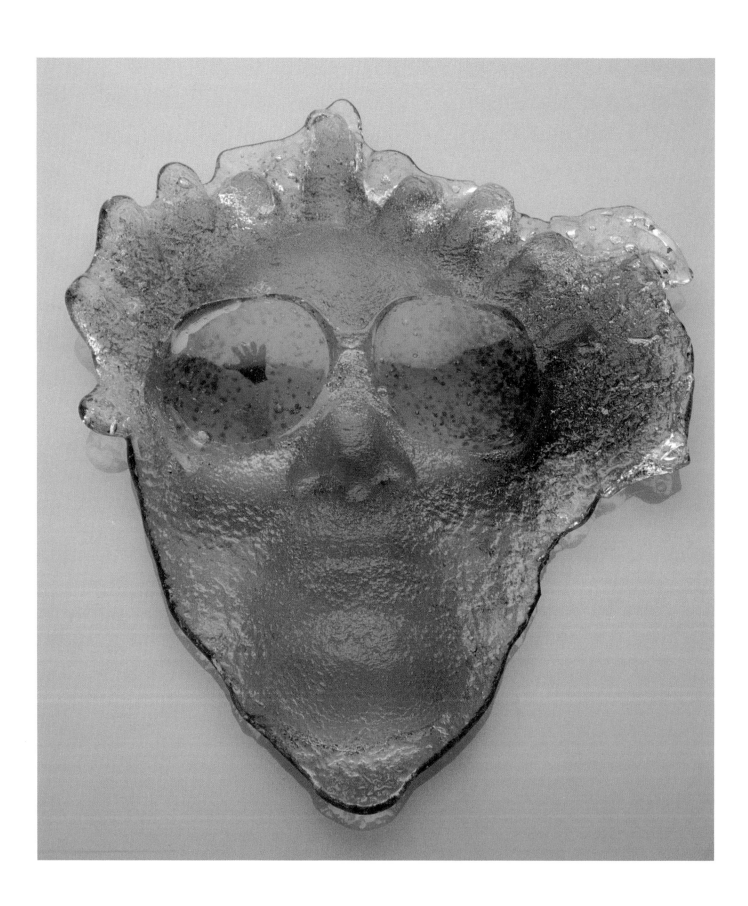

CHAPTER SIX
Metal, Clay & Sand Molds

METAL, CLAY AND SAND MOLDS

Metal molds, clay molds, and sand molds all have varied applications in hot glass techniques. With the exception of sand molds, these diverse forms can be purchased or adapted from other uses, although the individual artist may, at times, prefer to fabricate a mold to his own design.

Stainless steel and mild steel slump molds, cast iron and cast brass press molds, found-object metal molds of all kinds, and steel and copper wire sagging molds are general categories of metal molds. The properties of the mold material give each mold individual characteristics. Metal slumping and sagging molds accompany the glass through all stages of the firing cycle; casting and pressing molds do not.

Metal molds are excellent for slumping because they absorb and give up heat at a rate similar to that of the glass, as they accompany the glass through the total process. If they are made of iron or mild steel, metal molds tend to oxidize, causing the surface to flake off. Stainless steel is superior to other metals since it does not oxidize. Aluminum is not acceptable as a mold material due to its low melting temperature. Cast iron usually requires the use of a foundry for forming, but this is not as difficult as it may seem. Cast iron or cast brass are used in pressing and casting many similar objects.

It is tempting to think in terms of making your own molds and, as in any craft endeavor, great satisfaction may be gained by maintaining control of the entire process. Making metal molds is time consuming, but the expended effort is well worth the creative venture. The ability to make molds of any kind allows greater freedom in producing individual and unique glass forms. However, it is important to remember that the use of the mold (rather than its creation) is the primary consideration.

STEEL

Stainless steel molds are undoubtedly the most serviceable of all the steel molds. They have a highly polished surface and are not susceptible to cracking due to thermal shock or careless handling. Stainless steel mixing bowls are commonly used as slump molds, because they are inexpensive and readily available. They can be re-formed using various techniques (such as hitting with a hammer on an anvil, placed in a metal break) or can be cut into various shapes with a saw or snips, and then reassembled. Other stainless steel shapes such as pans and lids, flat sheet and bars and rods are also readily available. Stainless steel is not easily worked, in that it does not cut easily with a saw. Since it does not oxidize, an oxy-aceteline torch will not cut it as it will mild steel. Cutting thin stainless steel with tin snips or a fine tooth hack saw seems to work best in home studios. Cut pieces can be put together with stainless steel screws or rivets. Thin, flat sheet can be bent easily and scrap of many kinds can easily be purchased from sheet metal fabrication shops.

Mild steel is more easily formed, cuts easily, and can be welded or brazed together with relative ease. Mild steel has one drawback: it oxidizes. This oxidation is greatly increased during the firing process and the metal spalls from the surface after a few firings. Even if liberal coats of shelf primer are used, the result is a very rough surface.

Various stainless steel forms used as molds.

Opposite page: "Mental Digits", 10"x12", sand mold casting, Linda Ellis Andrews.

Cast resin mold, gray cast iron mold, and a cast fish taken from the cast iron mold.

Even though the slow breakdown of mild steel causes a few problems, I use it quite often. Angle Iron, pipe sections and found metal forms all have their application for the molding process. Every junk yard, second hand store, or welding shop has hundreds of shapes available at very low prices. Angle iron and pipe (use black pipe, not galvanized pipe) can be arranged, covered with fiber paper, and used as containment molds or slump molds. Mild steel is so inexpensive, that when oxidation creates too rough a surface and becomes a problem, nothing is lost by replacing the steel.

Cover with shelf primer all stainless or mild steel surfaces that will come in contact with the glass. Mix the primer twice as thick as usual, approximately 2 1/2 parts water to 1 part primer. Heat the mold to 200°-250°F in your kiln, and then apply one or two coats of the thick primer solution. Any further application will not make the primer thicker and will be likely to remove the good coating you have achieved. Because the metal is not porous, the water is not absorbed and must evaporate. The primer may appear thin, but that is to be expected.

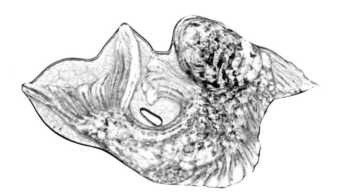

Metal, Clay & Sand Molds

CAST IRON AND BRASS

Cast iron and cast brass molds may be used as slump molds, but it is not common practice. The usual application for cast iron and brass molds is for casting or press casting glass. In the casting process metal molds do not accompany the glass into the kiln. The mold is usually cold or warm (below 700°F), and the objects are removed and placed in the kiln for annealing. Generally, no mold release is necessary; the glass chills next to the warm metal surface and does not stick. A simple one-piece mold is flipped over and the glass object releases and falls onto the metal casting table, chilling the flat top side. The cast piece is then carried to the kiln for annealing, on a charred wood paddle or other suitable material. Press molding is very similar to casting except that pressure is applied to the glass surface after it is in the metal mold. This can be as simple as a flat metal plate or involve a second form containing a design.

Gray cast iron molds used for multiple pressings.

A very ingenious cast iron mold system was made by one of my students. Curtis purchased two small cast iron frying pans, one approximately 1" smaller than the other. The smaller frying pan was ground on the bottom side to remove the name plate and small foot that had been molded there. He then ground a shallow incised design of flowers and leaves into the bottom of the small frying pan, using a flexible shaft grinder and carborundum wheels. Hot glass was poured into the larger frying pan and the second was placed on top and pressed down, forming a shallow bowl. Holding both handles, both frying pans were flipped over. The larger cast iron pan was removed, and then, with the glass draped over the smaller pan, he flipped it over onto a wood paddle to carry it to the annealing over. Within half an hour he had emptied six crucibles and made six bowls. Each had individual character and were marbled in various cathedral colors. Ingenious Curtis!

Simple cast iron molds are not difficult or expensive to have made for you. Briefly, you make a clay or wax model without undercuts, cast resin or Ultracal 30 gypsum cement over the model, remove the model, and shape the resin to the size and smoothness you want in your cast iron mold. The mold may be taken to a foundry to be cast in gray cast iron. Pattern resin can be found at foundry suppliers. When you call on a foundry don't mention that you are an artist, just ask for a "one off" casting: one copy, the first one. Foundry people are artists, too. If you say you are an artist they will want to do an exact replica of your pattern and may cast your pattern four or five times to achieve results with all the detail. This will cost you four or five times what you need to spend. Expect to pay for your casting by the pound, plus set up fee: approximately $35 to $50 dollars for a mold the size of your open hand.

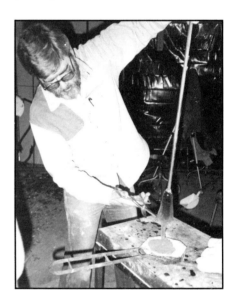

Filling a cast iron mold with glass trailed from a punty rod.

The mold should be finished with small grinding tools and carborundum sand paper. It's necessary to leave approximately 3/4 " of metal around the model impression and about 3/8" of metal on the bottom. Holes can be drilled and taped with threads so that you can add a handle.

Stamps for pressing designs into hot glass. Left to right they are made from wood, carbon, and metal.

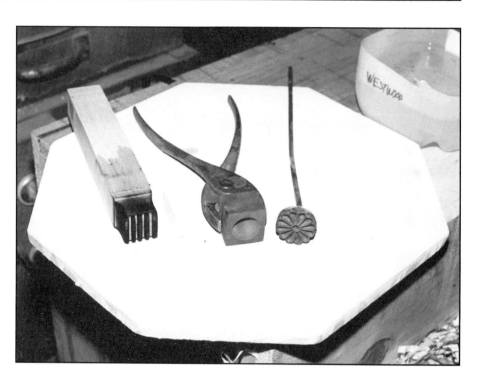

Sliding fired glass squares (at least 1/4" thick), ready for stamping at 1600°F onto a metal paddle from the top shelf of a top fired kiln.

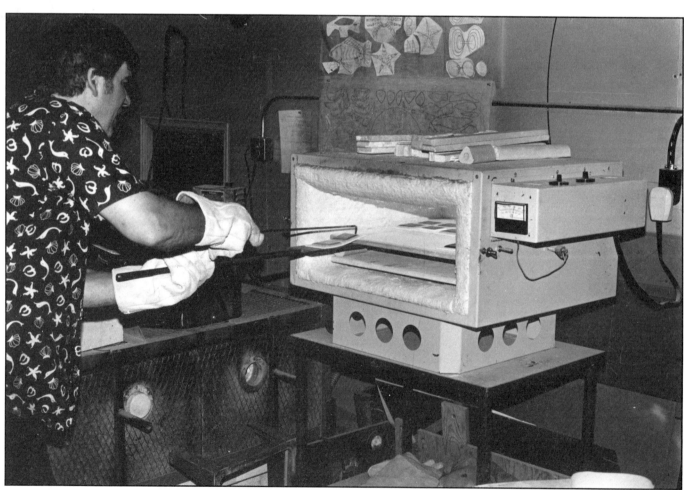

Metal, Clay & Sand Molds

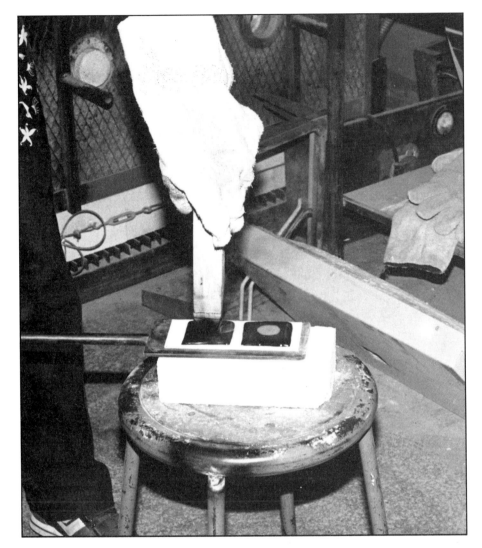

Pressing the design into the hot glass while it is still on the metal paddle.

The pressed tiles.

The imprinted tiles are placed on the bottom shelf of the hot kiln for annealing. The bottom shelf maintains a lower temperature because the kiln is fired from the top.

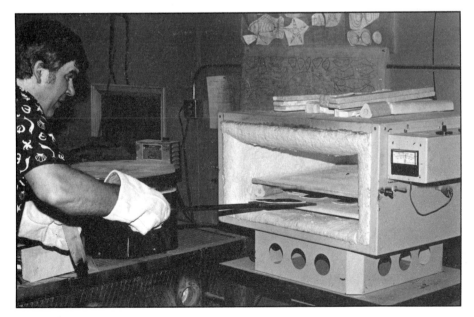

Pressing clay into flat slabs with the palm of the hand. In the foreground are various cut designs to be used for drop out molds.

CLAY MOLDS AND MODELS

Clay is one of the most versatile of all moldable materials. It can be slip-cast, rolled into slabs, thrown on a potter's wheel, or hand-formed. Stoneware clays form porous and very durable molds, when fired to bisque temperatures, 1600° to 1800°F. With proper care, bisque stoneware molds are long lasting. One disadvantage of clay molds is the 12% to 15% shrinkage from wet size to fired size. Clay molds also thermal shock readily, because of a change in volume of the silica particles within the clay body. This is known as quartz inversion, and occurs during heating and cooling cycles at approximately 1050°F. To avoid cracking clay molds, do not place hot glass on a cold mold, and do not crash cool your kiln by venting below 1300°F.

Clay molds are not porous enough to let air trapped between the glass and the mold escape. Before bisque firing your molds, drill air vents in the deepest cavities. For bowl molds, drill holes in the bottom sides, not the bottom center, since that is where the glass will touch first when slumping, cutting off the air vent.

Metal, Clay & Sand Molds

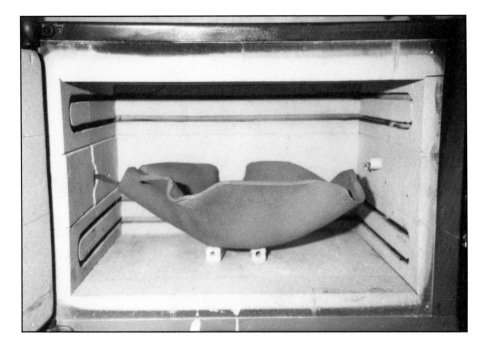

Slab clay draped over model, air dried for several days, the finally dried in the kiln.

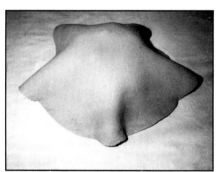

FORMING CLAY MOLDS AND MODELS

Coil building is one of the easiest methods of making clay molds. Any pliable, moist, stoneware clay may be used for this method. It is best to have a clear idea of the shape of the intended mold before starting. It is advisable to make a full size profile drawing before starting, then make a template or pattern from cardboard to check the contour as the mold is being built. After applying two or three coils, weld the coils together, and check the contours with the template as the work progresses. After all the coils have been applied, shape by scraping and sponging the surface. Any surface texture or incising is done while the clay is moist. Air drying for three to five hours will allow the clay to harden to "leather" hard. Detail should be applied at this stage.

The slab method of working clay requires forming the clay into pliable sheets of even thickness before starting construction. Place the clay on a damp cloth between two boards of equal thickness. The clay is formed by rolling a piece of pipe over the clay using the boards as guides. The thickness of the boards determines the thickness of the slab. Clay slabs can also be formed by pounding the clay flat between the boards, then slicing it with a wire held taut and pulled along the boards. I suggest thicknesses of 1/2" or more for substantial molds.

Edges to be joined, "welded", should be scratched and moistened before pressing together. Joints should be reinforced with small coils placed along seams and smoothed with fingers or tools.

Slabs can be draped over plaster, bisque pottery, or moist clay forms covered with a damp cloth. When the clay has stiffened enough to hold its shape (about three to five hours), it should be removed from the form. To prevent cracking of clay objects allow them to dry slowly. Cover them loosely with plastic for two or three days, then remove plastic for two more days. Fire on low setting for five hours before taking to 1800°F.

Metal, Clay & Sand Molds

SAND MOLDS

Sand casting glass is most often visualized as a process associated with a big glass furnace where glass is ladled into a large mold cavity by sweat-encrusted, strong men. But sand molding can be a small studio practice as a step in making plaster models, for use directly as a slump mold, for crucible casting, or as a mold for crucible pouring. Sand is easily formed, inexpensive, and a very immediate material to work with.

Green sand is the name given to any sand mixture that is made with a combination of different meshes of sand, mixed with water and clay as a binder. The clay holds the sand grains together to make it formable. Green

Sand cast glass mask by Linda Ellis Andrews.

Pressing Linda's face directly into the sand to make the impression to be used as a mold.

Eyeglasses are pressed into the face impression.

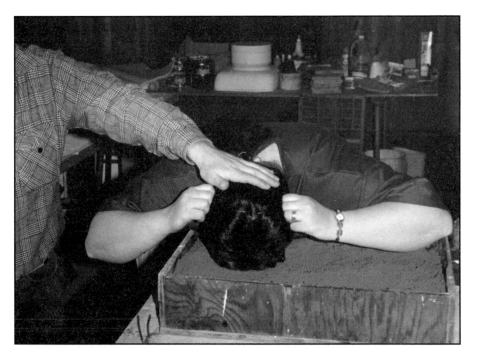

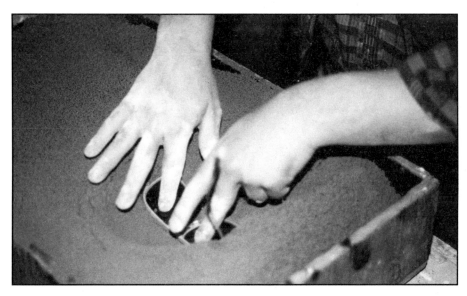

Carbon is applied to the surface of the sand with an acetylene torch. The carbon will act as a separator between the glass and the sand.

sand gets its strength and valuable forming properties from the controlled mixture of the sand size particles. Olivine sand, mined in Washington state, is a common foundry sand made from minerals that have a low thermal expansion, thereby making it an excellent sand for glass casting. It is not a silica sand.

Olivine "120" and "180" do not refer to mesh size. They are the names given to a sand of a specific combination of mesh sizes. Olivine 120 has 17% 70 mesh, 28% 100 mesh, 26% 140 mesh, 20% 200 mesh, 5% 270 mesh, and traces of other mesh sizes. When mixed with Olivine 180 (made of other mesh sizes), it creates a mixture that has great packing properties and will therefore take a very fine impression.

A GREEN SAND MIX
20 lbs. Olivine 120
20 lbs. Olivine 180
2.8 lbs. Bentonite (7%)
2+ lbs. Water

Mix sand and bentonite clay very thoroughly before adding water, straining it through a coarse screen to insure a homogeneous mix. If a large batch is made, store moist sand in a plastic garbage can. Keep separate some sand mix without water added in case sand gets too moist and must be dried. Approximately one gallon of water will moisten 100 lbs. of mix.

CASTING IN SAND MOLDS

Construct two wooden boxes with sides seven or eight inches high. Make a frame the same size as the box opening to hold a piece of window screen. Place the screen over one of the boxes and force the sand through the screen until the box is full. This process is called riddling. Riddling sand can be tedious, especially if the sand is too wet or the screen too small. Having two screen frames, one with 1/4" mesh and the other with window screen mesh, is advisable if a lot of sand is to be mixed. An initial screening through the larger mesh will save time and effort. Some students riddle while they riddle, posing questions to exercise their ingenuity, hoping someone will discover their meaning.

When filling the casting box, use sand forced through the course screen for the bottom half of the box, then finish with sand screened through the fine screen. Use a rubber scraper or wear a leather glove to keep from wearing off your finger tips. Sand can be placed in the four corners and along the edges, before screening, to save time. If sand piles up in the middle of the box, gently move it to the sides, being careful not to compress the sand. When the sand depth is twice that of the object to be pressed into it, the riddling is complete.

Any non-porous, non-sticky object or combination of objects may be used as a pattern to press into the sand. Wax and leather-hard, clay models should have graphite powder rubbed over their surfaces to keep the damp sand from sticking to them. Graphite powder used for lubricating locks can be purchased at the hardware store. The sand impression may have under-

Metal, Clay & Sand Molds

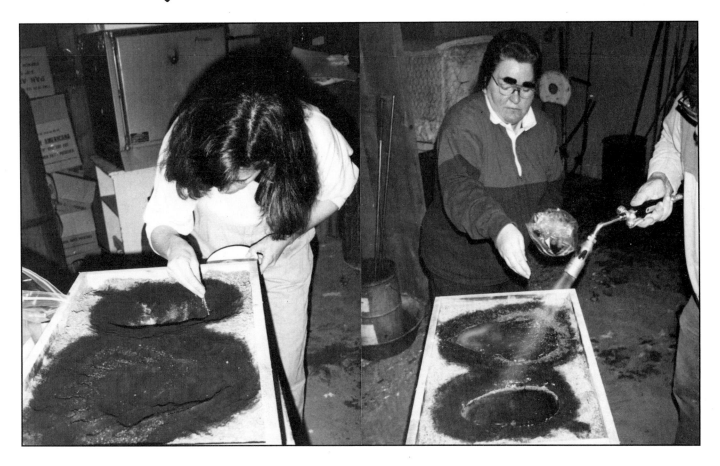

cuts, "back draft". This is accomplished by pressing objects into the sand at different angles. If small grains of sand fall into the impression they can be picked out with a wet Q-tip cotton swab lightly touched to the sand particles; they will stick to the wet cotton. Having good handles on patterns will help keep sand from falling into the mold cavity, since most flaws are caused by difficulty in removing the press pattern.

Visualizing the object that will be formed in the negative space is difficult for some people. Rather than committing a full crucible of glass to your first sand molds, I suggest pouring plaster into the sand cavity. After 15 to 20 minutes, the plaster object is set and can be removed. Sand that sticks to the plaster can be removed with a brush and water. The sand mold can't be used again, without rescreening, but the screening process is relatively easy and fast. I suggest having two mold boxes so the sand can be screened into the second box without transferring it into another container. Be sure to remove any plaster overpour.

Plaster objects made in this way may be used as a pressing pattern if they do not have undercuts. Remove any back draft and rework the surface with scrapers and clay working tools. Insert pieces of bent coat hangers in the back of the plaster casting, shortly after pouring, to make handles for easy removal. Add texture or radically change the plaster form if you wish. This is often desired if your face was the original model pushed into the sand mold. Most students want to change their nose, because they seem to

Frit, enamels, and other design elements are added after the carbon has been applied to the sand. Linda adds design elements of copper and glass between glass pours, as theleggii glass edges are kept hot with a hand torch.

Boyce ladles the second glass pour over design elements causing them to become interior in the finished sand casting.

flatten out or get pushed to the side when pressed in the sand. Others change everything, claiming that they want to practice being a plastic surgeon.

Plaster patterns can be coated with stearine and kerosene while still moist (uncured) or dried in a warm oven, then sprayed with fast drying lacquer. The intent is to have a dry, smooth, non-porous surface to press into the sand.

Metal, Clay & Sand Molds

Riddle the sand once again. If the prepared sand is not going to be used right away, cover it with plastic to keep the surface from drying out. If the surface does dry out or if the sand has been sitting for more than an hour, spray the surface with water, using a pump spray bottle. Test the ability of the sand to make a good, detailed impression by pushing a small object in the corner of the mold box. When pressing deep or large shapes (such as a face), scoop out the center of the sand pile with a spoon or small trowel, then sift approximately 1/2" of fresh moist sand over the hole. When pressing more than one form into the sand, press the deepest first, then add texture with smaller objects.

After the sand impression is satisfactory, a release of carbon black or other suitable material should be applied over the surface. This is not absolutely necessary, especially if a coarse sand texture is desired. Sand that sticks to the glass surface can be removed with a wire brush or a sand blaster, although some sand will roll into the surface and cannot be removed without deep blasting.

Most commonly, carbon black is applied to the surface of the sand with an oxy-acetylene torch. The torch is turned on and adjusted to a standard cutting or welding flame then the oxygen is turned almost off. This will cause billows of carbon black to pour from the torch head. By holding the torch 4 or 5 inches from the sand and moving the head back and forth across the surface, carbon is applied to the surface. The surface should be completely covered with no light areas showing. Too much carbon is never a problem. At this stage, colored frit, stringers or enamels can be applied over the carbon layer. Now the sand mold is ready for glass.

Not all studios have an oxy-acetylene torch, so other surface treatments can be used. Lamp black (carbon black) mixed with denatured alcohol can be sprayed over the surface with a hand pump sprayer. After an even coat has been applied, light the alcohol. This is exciting. Graphite powder, generally used for lock lubrication, comes in small plastic tubes, and can be dusted over the surface, if done very carefully. Hold the tube horizontally, seven to ten inches above the sand, and pump the tube between thumb and forefinger. This pumping action should mist the graphite powder over the sand impression. Try not to let it pile up in one place. Graphite powder can also be sifted over surface with an enamel sifter.

Deep or large sand molds should be vented. In general, when more than 10 lbs. of glass are going to be poured into the mold, holes should be poked into the sand, starting 1 1/2" to 2" from the outside edge, angling under the object, and approximately 2 inches apart. A pencil or welding rod can be used to make the vent holes. Venting sand may keep bubbles caused by steam from rising through the glass. Bubbles that do form should be heated with the torch; they will rise to the top and pop.

Have all necessary tools accessible before pouring glass. A pair of diamond shears should be handy to cut the last trail of glass from the crucible. Keep thin areas of the casting, which cool faster, hot by heating with a hand-held torch. Use a small trowel to dig sand away from the sides, exposing thick areas as soon as possible. The idea is to have the casting cool evenly.

The glass is removed from the sand using wooden and metal jacks. About one inch of sand adheres to the glass, to be removed after the annealing process.

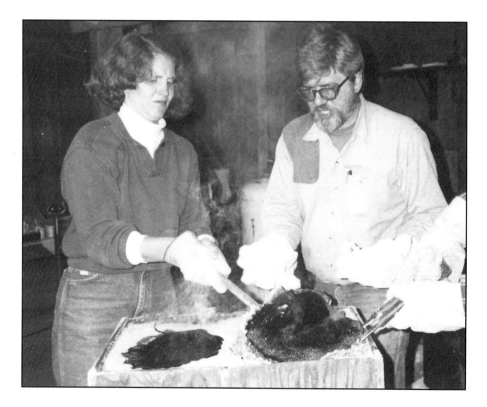

Remove the glass from the sand while it is still showing a dull red color. Most of the sand will fall off in the casting box. Use charred green wood sticks to lift the cast piece onto a plywood board and take it immediately to the kiln for annealing. The concept is to get the casting to the annealing oven as evenly heated as possible and as hot as possible, without having it deform. Differential temperature causes stress. If handling the cast object causes chilling in some areas, it will have to "soak" at an elevated temperature (1100° to 1150°F) for an additional two to three hours and become equalized before the annealing process can start.

Don't handle the glass casting with cold gloves, and don't heat your gloves on top of the kiln. Gloves left on top of the kiln are often forgotten and catch on fire. For pieces that can't be handled with charred sticks, heat two pieces of 1" fiber blanket on top of the kiln. Support glass on warm fiber blanket for transporting to the annealing oven.

After casting into the sand, remove any small glass threads or other foreign material. Spray the hot sand with enough water to dampen dry areas and riddle into second box. Sand may be used 5 to 10 times if remixed well. In time, the bentonite in the sand will lose its bonding properties. Add bentonite to increase bonding strength. Sand that is too wet will stick to press patterns and cause steam bubbles in the cast glass. Add just enough water to the sand so that when a handful is squeezed, the palm print can be seen and the lump stays in one piece.

Metal, Clay & Sand Molds

The surface of the sand cast piece can be polished using a diamond sanding pad attached to a Dremel tool.

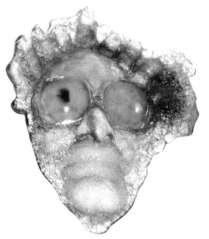

CASTING STUDIO CHECKLIST

Casting glass into sand, into plaster castables, or into metal molds is a process that works best if there are two or three people forming a team. Timing is important when pouring any quantity of molten glass. Talk through, perhaps walk through, the process with your helper at the beginning of each casting session. Know who has responsibility for each process. This is done by professionals *every* time they cast glass. *Have non-participants stay in one place.*

PROCESS REVIEW

1. Prepare molds and have them covered with plastic.
2. Have glass in crucibles holding at 2000° to 2100°F.
3. Check self and surrounding area for flammable materials. Have a bucket of water available.
4. Make a diagram chart above the kiln showing what glass goes in what mold.
5. Have one person in charge of turning the kiln off before removing crucible. Make sure they know to *turn kiln back on* if more than one crucible is being cast, especially if kiln is electric. This is not as much a problem with gas kilns.
6. Move crucible to bricks on casting table and reposition tongs for secure grip on crucible before pouring. Don't hurry, there's plenty of time.
7. Pour glass, moving the crucible so glass will flow into extended areas of the mold.
8. Cut glass trail from crucible with tin snips or diamond shears while over center of the casting. Don't try and remove overpour or threads that are unwanted until after glass is annealed.
9. Heat coolest areas of casting with torch until piece is ready to be moved to annealing oven.
10. There is plenty of time to pour more than one crucible of glass into a mold. Addition of pre-fused glass pieces, frit, copper or stringer can be placed on top of the first pour before adding the second.
11. Transport the cast piece to annealing kiln while as hot as possible and with as little handling as possible.
12. It's O.K. to get excited, but *DON'T RUSH!*

CASTING MATERIALS CHECKLIST

1. Prepared casting sand or other molds.
2. Crucibles and crucible tongs (fireplace tongs).
3. Water bucket.
4. Glass shears.
5. Torch.
6. Gloves and 1" fiber blanket.
7. Green wood sticks and plywood shelf.
8. Safety glasses.
9. Pump spray bottle/water.

Metal, Clay & Sand Molds

KILN FORMING GLASS IN A SAND MOLD

Olivine sand mix can be used in the kiln forming process, as well as for open-faced casting. Sand preparation and riddling procedures are the same for both. A sand containment ring that can accompany the sand mold through the kiln firing process is necessary. Thrown or hand built ceramic rings, steel well casing or 12" pipe rings and many other refractory materials will work as a sand containment ring.

Place the containment ring on a kiln shelf of the same size. In other words, don't cover just part of the shelf with a sand mold or the shelf is likely to break during firing. Fill the ring with sand, using the procedures detailed in the sand mold chapter. Press the pattern into the sand, remove the pattern, and then spray with colloidal alumina or mold hardener, using a pump sprayer.

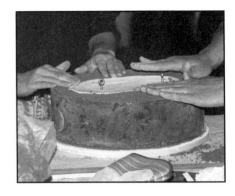

Compressing the sand between the plaster pattern and the containment ring.

After the plaster pattern is removed, the sand is sprayed with mold hardener.

Frit and small pieces of glass are added to the impressed sand mold after the mold has dried in the kiln.

Dry the mold in the kiln at 400°-500°F. This will take from one to three hours, depending on the thickness of the sand, the amount of moisture present, and the mold depth. Not as much sand depth or surround is needed with a kiln fired sand mold, because the sand and the imprint are held in place by a refractory ring. After the mold is dry, spray it with shelf primer and allow it to dry before placing glass over or into mold.

Sand molds provide a quick way to make a mask or impression of your face. Face molds may be filled with frit, or layed up with various sizes of broken pieces. One advantage of using overlapping pieces of sheet glass is that the glass does not have to fill the mold, so an extraordinarily long anneal is not necessary. Fill deep impressions with small pieces of cullet and then place two layers of glass over the mold cavity and fire to 1450°F. The result will be a hollow impression.

Metal, Clay & Sand Molds

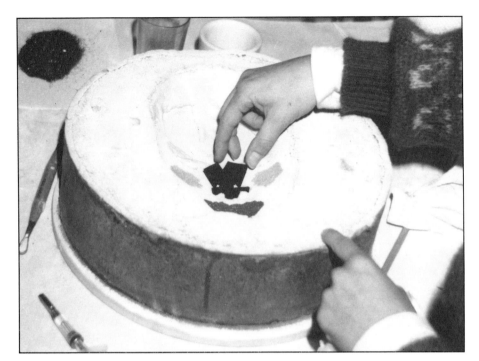

The hardened sand surface may be sprayed with shelf primer before the mold is filled with glass.

Masks made in the kiln fired sand mold method by Joan Malone.

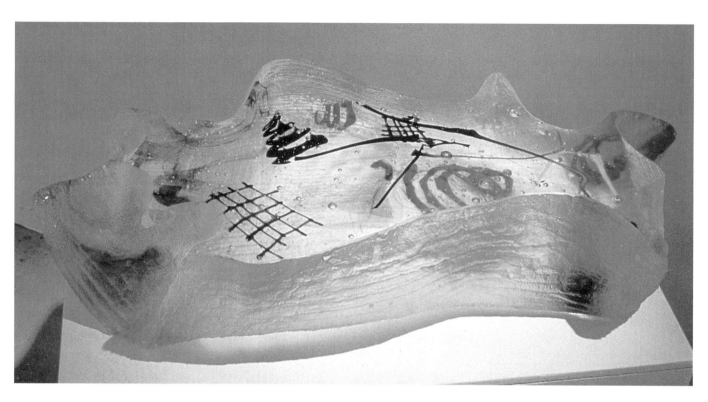

CHAPTER SEVEN
Fiber Molds

When alumina fiber material is saturated with fiber mold hardener and heated in an oven or kiln, it becomes rigid and forms a hard shell surface. This combination of materials is referred to as *moldable fiber blanket*. It is used for making original glass slumping molds or for reproducing shapes of objects that are not refractory enough to be used as glass slumping molds. Plastic, aluminum, and glass are examples of materials that cannot be fired in the kiln to the temperatures used in glass slumping. However, many objects made of these materials can provide interesting forms for slumping molds. The reproduction of a bent glass lamp panel is a good application for moldable fiber blanket, since the fiber shrinks very little (approximately 1%) and a direct mold may be taken from an existing panel.

Alumina fiber products are readily available through refractory supply companies and fusing supply craft outlets. The most common fiber products used by glass enthusiasts are fiber paper, fiber blanket, fiber board, and chopped bulk fiber. These fiber products are used to insulate kilns and glory holes, as bas-relief molds, as modifiers to castables, and, in conjunction with the proper binders, to make rigid fiber molds.

Either colloidal silica or colloidal alumina is used as a binder and hardener with any fiber product to make moldable fiber materials. However, there are many prepared (pre-saturated) moldable fiber products on the

Fiber paper is placed over steel pipe sections of various sizes. Molds made by this method are infinitely variable. Shapes can be changed by adjusting the distance between the supporting elements..

Opposite page: Serie: " 'L'Avesnois' Water Ice Dreaming", 16.5 x 70 x 32 cm, cast glass, David Ruth.

Serie: " 'L'Avesnois' Menhir Marker", 126.5 x 25 x 19 cm, cast glass, David Ruth.

Fiber products.

market, which may be purchased from glass fusing suppliers. Moist Pack, Wet Felt, and Zircar fibrous ceramics are some trade names of these products. The process of making a fiber moldable is fairly simple and may be more economical than buying commercially prepared products, especially if large molds are desired.

Fiber blanket is available in 1/4", 1/2" and 1" thicknesses and in densities of 4, 6, and 8 pounds. The 6-pound density, 1" thick blanket can be used for all studio projects, from making molds to lining a crucible furnace or constructing a glory hole. One inch fiber blanket can be separated into one-half or one-quarter its original thickness by simply pulling it apart. Therefore, one roll of 1" fiber can be a wise purchase, if you plan on making a lot of molds or building equipment.

Fiber Molds

MAKING A MOLD

After choosing the shape to be made into a slump mold and gathering the few materials necessary, you are ready to start the simple mold making process.

Cut the fiber blanket roughly to the size of the chosen form. This will eliminate waste of fiber and mold hardener. If you happen to wet more fiber than is necessary, you may store the excess fiber in a plastic bag, and, as long as it is not exposed to air or heat, it will not dry out or harden. You can cut the fiber with scissors or a razor knife, but don't use your best dressmaker's shears, because the fiber has a tendency to dull the scissors.

After cutting the blanket to approximately the size of the form over which it is to be molded, saturate the blanket with mold hardener. I suggest a shallow pan like a large cookie sheet, or sheet plastic supported around the edges by wood risers, to catch the liquid. In this way, the mold hardener will pool so that the blanket will become saturated.

Pour the mold hardener over the blanket and press until the mold hardener has thoroughly soaked the blanket. A piece of two or three inch plastic pipe works as a good rolling pin. It takes approximately 1 1/2 pints of mold hardener solution to saturate one square foot of 1", 6 lb. density fiber blanket. The fiber blanket will compress, as you wet it, into half of its original thickness. Starting with one inch blanket, after thorough soaking, you will end up with half inch moldable blanket. If you start with 1/2" blanket, you will end up with 1/4". If the mold is to be used more than a few times, start with one inch fiber blanket; it makes a much more substantial mold. For reproducing bent lamp pieces, thinner material is adequate.

Rolling fiber mold hardener liquid out of saturated fiber blanket.

This intricate mold was constructed with fiber paper, fiber board, and carved plaster by Mike Dupille. The elements were cut, bent, and held in place with white glue, then sprayed with mold hardener.

Please note that fiber mold hardener is not a glue. You must use one piece of blanket. Two pieces in a mold will not stick together.

After the blanket is thoroughly saturated, squeeze out, by rolling with rolling pin, all of the excess mold hardener. This can be poured back into a container and used again. The fiber blanket is now moldable and ready to place on the form.

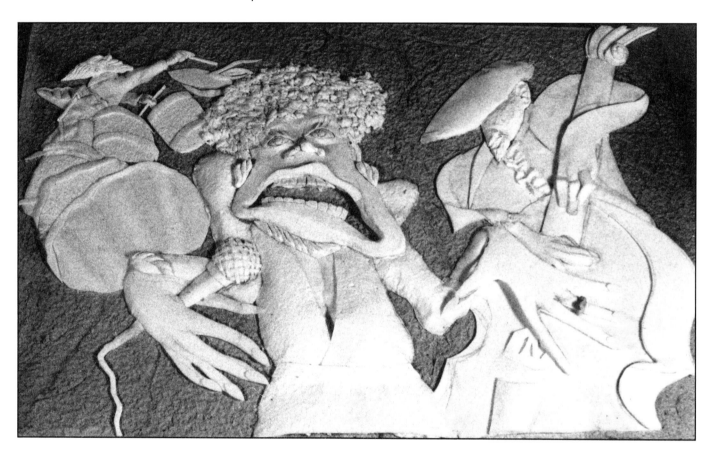

It's important to put a mold release on the form to be covered. There are many materials that can be used as a mold separator. Spray-on cooking oils or Vaseline, applied in a thin coat, work very well. Aluminum foil or plastic wrap also works as a separator on simple, uncomplicated molds.

The fiber is placed over the mold and slowly worked or pushed into place. It should conform very closely to the mold, but at first the fiber blanket will have a tendency to spring away. Work and press until it gets very close to the mold. When fully prepared, it should cover the form, although in some places the fiber blanket may stretch a little and become thinner. You cannot overwork the material by pushing and shoving it into place.

After the blanket has been pressed over the mold, it is trimmed with a razor knife. Small pieces can be saved for other projects. After the excess mold material is cut off, compress it again with the fingers all around the mold edge, pushing it down and holding it in place.

Fiber Molds

At this point the mold is ready to dry and harden in a regular oven or a kiln. A hair dryer or hot air gun can be used to dry areas of the mold so that they will hold their shape. Unaided air drying will take about two days. It is okay to dry the mold by placing it in a preheated kiln. With the kiln set on low and the lid cracked open two inches, it will reach about 250°-350°F. At this temperature, it will take approximately two hours to drive off the moisture and harden. After one hour the fiber is semi-hard and will hold its shape. At this time, it is taken out of the kiln, removed from the form, and the plastic wrap is removed. The mold is turned over and put back into the kiln so that it can dry on the inside surface.

During a kiln drying, water will be driven off in the form of steam, so it is important to have the kiln lid open to allow the steam to escape the kiln. This facilitates faster drying. Some mold hardeners contain glycol, added to colloidal silica to keep it from freezing. When it is driven off with the steam during a kiln drying, it has an acrid smell and can be harmful in large amounts, *so work in a well-ventilated studio.*

After the mold has dried out, it can be further worked to clean up surfaces and edges. Edges are trimmed with a razor knife and surfaces sanded where the glass will touch during slumping. You will notice, after sanding, that some parts of the mold surface are softer and become difficult to sand. When this occurs, very liberal coats of mold hardener should be applied with a brush to the working surface of the mold (the surface that the glass will be touching). Then the mold is put back in the kiln and fired to 350°F.

If a very smooth surface is required, sandable mold coat is applied (a formula and process for making this paste is discussed later in this chapter). Very satisfactory results can be obtained by applying two or three coats of sandable mold coat, sanding down to the surface of the moldable blanket after each coat. You may note that, when sanding through the original mold surface, a sixteenth of an inch or so, the mold surface again begins to get soft. This is natural and the way that mold hardener works—it only hardens the surface of the fiber. In most instances, three coats of mold coat will have to be applied and sanded before a truly fine surface is acquired. When the mold is satisfactory, it should be fired to 1300°F before it is used. This will make the mold a lot harder than it was when only fired to 350°F or 400°F. When sanding, wear an approved mask and clean up all dust when work is completed. Repeated or prolonged breathing of alumina-silicate should be avoided.

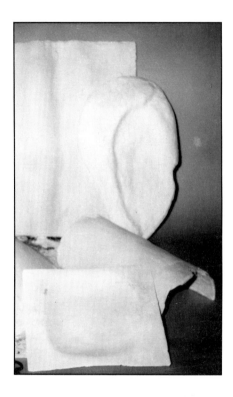

Preformed fiber molds.

FIBER MOLDS DO NOT THERMAL SHOCK

FIBER MOLD APPLICATIONS

Ease of forming, low shrinkage, and durability at high temperatures are advantages of fiber molds.

Reproduction of bent lamp forms for repairs is one place where fiber is at its best. Since its shrinkage from wet to dry is so low, a mold can be made from a surviving bent lamp piece, and used as the pattern. When using glass as the pattern, the mold may be dried in the kiln, if care is taken not to thermal shock the glass.

Large slump molds the full size of the kiln can be made by forming the fiber blanket over vermiculite in the kiln. Vermiculite is poured into the bottom of a top fired fusing kiln in sufficient quantities to form the desired shape. Water is added to the vermiculite to make it moldable. After the desired shape is formed, prepared fiber blanket is placed over the vermiculite and pressed into the contours.

The fiber mold is then fired to approximately 400°F. This will take two hours, because the water in the vermiculite and in the mold hardener must evaporate. It is important to have the kiln door cracked and to work in a well ventilated studio.

After all water has been evaporated from the mold, cool the kiln and remove the vermiculite. Put one more coat of mold hardener on the mold and fire to 1300°F. The fiber mold will now be rigid and ready for a sanding mold coat. Three to four applications of mold coat, dried and sanded between applications, will give a very smooth mold surface. Molds three feet to four feet and larger may be made in this manner. The large molds should be supported with pieces of insulation brick or kiln shelf supports during a slump firing.

Fiber molds should last for hundreds of firings. Fiber product molds are not lost due to thermal shock, a common cause of loss of many other kinds of molds. Fiber molds are more fragile than molds made of other materials, so it is necessary to handle large fiber molds carefully. Large molds can be reinforced or repaired by adding strips of fiber to the underside of the mold, gluing it on with Mold Mix 6, a Zircar product, or with Koawool Cement, a Babcock & Wilcox product. Alternatively, there is a fiber cement recipe given in this chapter.

Fiber molds may be used at higher than slumping temperatures. Most fiber blanket is rated at 2300°F or higher. Therefore, containment molds or casting molds may be made of fiber, if it is properly supported.

David Ruth has been constructing large casting molds in the bottom of his kiln. He supports the molds with bricks and pieces of kiln shelf, using the walls of his large, top fired kiln as the outer support.

David ladles glass from a furnace or pours it from a crucible, while holding the mold at approximately 1100°F. He alternates layers of poured molten glass and placed pre-made (hot) pieces of fused glass. This process may take three or four days, during which the kiln must be held at 1100°F. After the piece is complete, the entire casting, in the fiber mold, is raised to 1500°F. The glass levels and larger bubbles rise through the surface of the glass. Many of David's pieces approach 400 pounds in total glass weight. After a proper annealing cycle, sometimes as much as three weeks, the pieces are removed from the kiln and then ground and polished.

Making a fiber mold using vermiculite in the kiln.

Opposite page: "Cosmos Dreaming", 40" x 22" x 9", cast glass, David Ruth.

Fiber Molds

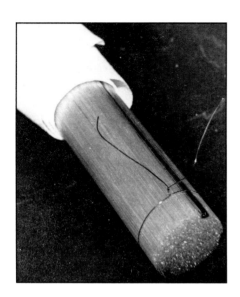

A stacked stringer design bundle is wrapped with fiber paper and then bound with wire.

Extra heavy duty aluminum foil covered with fiber paper.

Ceramic fibers, when held at an elevated temperature for long periods of time, may change in composition and become cristobalite, which is a short fiber and has qualities similar to asbestos. This may be encouraged by the loss of volatile fluxes from the molten glass. For this reason, it is safer to use fiber molds that have been used next to molten glass and at elevated temperatures only once. This is not a problem when using this type of mold as a slump mold.

FIRING FIBER MOLDS

The faster wet fiber molds are heated, the more risk there is of their deforming and losing their original shape. If a lot of steam is produced all at once, the steam will cause the fiber material to change shape. This is only a concern during the initial drying cycle.

Fiber molds do not thermal shock, so they can be used for various glass forming techniques where other types of molds would not work. It is possible to remove a piece of fused glass from the kiln at 1400°-1500°F, place it over a fiber mold, and place the mold and glass back in the kiln to slump and anneal. The surface of the fiber heats very rapidly and does not thermal shock the glass. Just under the rigid surface of the mold, the fiber is very open and porous. This allows any trapped air to escape through the mold. Only the deepest mold cavities need to have air vent holes drilled. Holes can easily be drilled after the fiber mold is complete.

Fiber blanket is an insulator and only the outer surface becomes dense in the hardening process. The finished mold retains its insulating properties. After glass has slumped into the mold, the glass surface in contact with the mold will cool much slower than the top surface. For this reason, it is necessary to anneal all glass objects slumped in fiber molds longer than when using clay or metal molds. I suggest firing down during the annealing process. The goal is to stop the temperature drop at 890°F and add heat slowly for approximately 30 minutes to reach 950°F, before resuming cooling. If the kiln has a controller, it can be set for an extra 30 to 50 minutes in the lower end of the annealing zone.

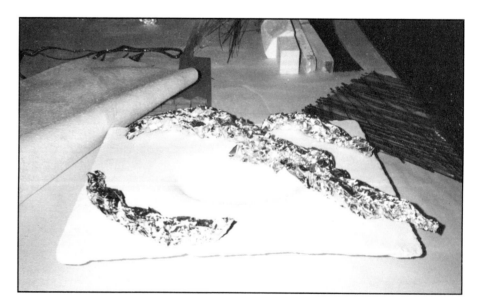

Fiber Molds

Fiber blanket supported on kiln furniture and a bisque fired ceramic vase.

A slumped glass form is made for both sides of the ceramic vase, then glued to it, using the vase as a supporting structure.

Fiber paper, glued in a ring to be used as a containment mold, will hold up to one-half inch of glass in place.

SANDABLE MOLD COAT

Mold coats for filling in rough areas in rigidized fiber products are manufactured by the same companies that make fiber blanket. Most are quite expensive and usually are packaged in one gallon and five gallon quantities. These products can be hard to find in smaller quantities, if they have not been repackaged by fusing product suppliers, due to their limited use and the dearth of published papers on their artistic uses.

A very suitable mold coat (filler) may be made in the studio if you have a blender. Scraps from fiber blanket or fiber paper (previously fired) may be recycled into many products. Chopped fiber, referred to in many recipes, may be made by pulling apart all scrap fiber into little pieces. These pinch-sized pieces can be used in place of the commercial product known as chopped fiber. If blended with sufficient liquid on HI speed for three or four minutes, the fiber pinches become very much like milled fiber.

MOLD COAT-S.F.
1 cup mold hardener (colloidal silica 15% soln.)
2 cups chopped fiber (approximately 30 grams)
1 cup silica
1 cup shelf primer

Place the mold hardener in blender. Slowly add chopped fiber. Blend at medium speed until all fiber is in solution. This may take three or four minutes. Pre-mix silica and shelf primer, then slowly add this mix to the fiber solution until the mixture reaches a creamy consistency. Mold Coat S.F. should have the consistency of thick clay slip. Store in an air-tight, covered container between uses. This material has low shrinkage and good adhesion properties.

Fiber Molds

Apply with a brush or your finger, until coating is even, approximately 1/8 to 3/16 inch thick. Air dry or place in a kiln at 200°-250°F. If the mold coat cracks, dry slower. Sand Mold Coat S.F. until 50% of the surface of the mold shows through. Apply successive coats and treat the same way until satisfied with mold surface.

FIBER ADHESIVES

Adhesives for gluing fiber blanket together are manufactured by all of the companies that manufacture fiber products. Most must be fired to 1600°F to achieve full bonding strength, at this temperature they develop a ceramic bond.

Kaowool Cement, made by Babcock and Wilcox, is an air-setting ceramic fiber cement for use with all fiber products. The cement sets at approximately 150°F, but should be fired to 1600°F to achieve full strength. Kaowool Cement is applied with a brush or spatula and is mainly used for joining ceramic fiber products to each other. Cera-Kote, by Manville Corp., is an adhesive *and* a mold coat, although as a surface coating it is hard to work with and is very expensive. Alumina Cement, made by Zircar, is an alumina adhesive designed to be used with alumina or alumina silica fibrous insulating products. It is available in one pint containers from Zircar.

All these products vary in cost because they are manufactured for various industrial uses. Most have much higher service strength and temperature ratings than necessary for use in glass molds. Making your own fiber adhesives may be somewhat time consuming, but having control of your total process may be less hassle in the long run.

FIBER GLUE
1 cup colloidal silica (20% soln.)
1 cup chopped fiber (approximately 30 grams)
30 grams milled fiber
1/2 cup alumina oxide
1 cup sodium silicate

Put colloidal silica in blender on high speed, then add chopped fiber. Blend until smooth. Add milled fiber and alumina oxide. Blend approximately one minute. Finally, add the sodium silicate. The mix will jell and it may be necessary to force the material into the mixing blades by pushing the Fiber Glue down the blender sides with a rubber spatula. Store covered in an airtight container.

Use approximately 1/8" of Fiber Glue between fiber blanket pieces to be bonded. Wet-to-wet or wet-to-pre-fired fiber material both bond well. This glue may discolor when fired. Fire to 300°F to bond and 1400°F for total strength. Fiber Glue does not thermal shock and may be fired fast.

ZIRCAR FIBER CERAMICS

Zircar is a company that came into being in response to the need for high-temperature fiber products for space travel and other applications affected by the highly technical research of modern times. They produced a side line of products specifically for glass and ceramic artists seeking to take advantage of advanced technology in their work. Luminar was the name they gave to this new venture and, for many of us working with contemporary fiber ceramics in conjunction with glass, they were luminous leaders. I use the past tense because Luminar is no longer in business, probably because the technical potential for artists exploiting the field of fiber ceramics is not well enough understood. Although Luminar is no longer in operation as a company, the Luminar product line is sold under the company name Zircar.

Zircar is an industry leader in the field of high-temperature, exotic refractories made with materials like zirconia, alumina, hafina and ceria. Zircar is a large professional, technical company accustomed to dealing with industry. But they do *have a heart* and a concern for the artist. You can deal with them directly and they will sell you Luminar (Zircar) products. Know what you want and try to be as professional as possible. They have a $25.00 handling charge for orders under $100.00.

I have tried to list and make comments on all of the Zircar products presently available that are useful to the glass artist. I have done this because, to my knowledge, there are no other products like Zircar fibrous ceramics on the market.

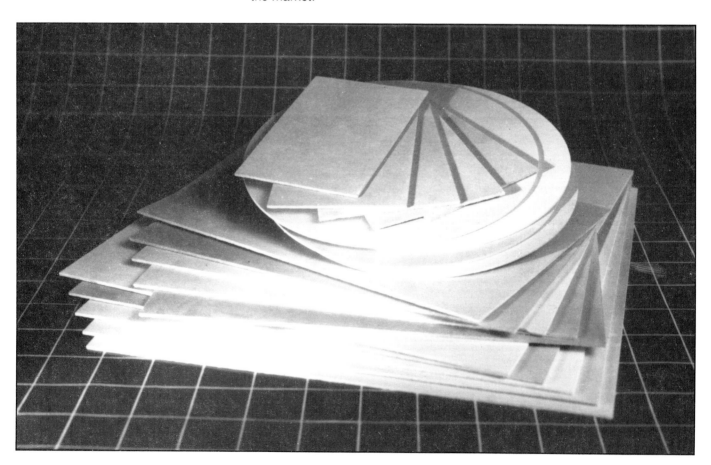

A selection of Luminar fiber paper products.

Fiber Molds

A mold constrtucted of Luminar LMS 80 moldable sheet. Although this mold appears fragile, it is very strong; damage might be done by rough handling or improper storage, but it will hold up well to the weight of the glass it is to hold.

LUMINAR MOLDABLE SHEET (LMS) is a high-alumina ceramic fiber material made into thin sheets. It has *excellent* strength for its thickness, and can be fired repeatedly to 1800°F or higher with no shrinkage. It is completely safe to handle and contains no asbestos.

LMS is supplied in dry form. It can be used directly in flat form as fusing shelves or it can be molded by wetting with water and shaping to any desired form. LMS becomes permanently rigid at 570°F. Joints can be made by pressing the two wet surfaces together; greater thicknesses can be made by laminating two or more thicknesses together. The surface can be carved to create texture. As long as LMS is not cured at temperatures over 300°F it can be rewetted to change the shape.

LMS is wetted by applying water with a brush or sponge across the entire surface, until the fiber sheet is saturated. It is allowed to soak for two or three minutes, then turned over and the process repeated. With thick sheets, the wetting process is repeated on both sides a second time. Moist sheet should feel like wet leather. If the sheet is not flexible enough, more water may be applied; if the sheet is overly moist, the excess water can be blotted out. Any excess sheet can be dried and reused.

When making joints, press wetted surfaces together firmly. Luminar Alumina Coat will create an excellent bond, if painted on the surfaces to be bonded.

When molding to an existing shape (pattern), use a mold release. Dried molds should be sanded to the final desired surface before being fired to high temperature. Complex shapes are made easy, because moldable sheet is thin, flexible, and sticks to itself. Slits can be made and edges overlapped; excess pieces can be used to reinforce thin areas.

Spraying a sand mold with colloidal alumina.

LMS comes in three thicknesses: .020 inches, .040 inches, and .080 inches, and in one size: 18" X 24". Special sizes can be ordered up to 3 ft. X 4 ft. I use one thickness, .080 inches (approximately 1/8 inch), for all molds, because I tend to work large and need the support for large pieces of glass. One-eighth inch of moldable sheet with overlapping joints will make a very substantial 24" diameter, 10" deep mold.

Specially ordered large sizes of LMS can be used for a large shelf, with no joints. Support the sheet with pieces of kiln shelf and stilts. Because the surface is so smooth and flat, I suggest covering the entire area with Luminar Ceramic Paper (LCP) or other fiber paper to avoid trapping air. LMS is a specially prepared version of Zircar's industrial grade known as RS type DD.

LUMINAR ALUMINA COAT (a rigidizer) is made with colloidal alumina. It has many uses described in this book. Because of its alumina base, it resists glass adhesion; glass does not stick to it, which it may when colloidal silica is used. Although this rigidizer can be used without shelf primer for one or two firings, I suggest always using a separating agent. Sodium vapors impregnate the surface of molds and shelves after a few firings, especially at high temperatures (1700°-1800°F) and the glass may stick.

Colloidal alumina has a higher tolerance to freezing than colloidal silica. This is one of the great advantages of colloidal alumina. The material may precipitate out of solution somewhat, upon freezing or after sitting for a period of time, but vigorous agitation will put most of the micron-sized particles back into suspension. Colloidal silica must have the addition of glycol if it is to be shipped in the winter months, because it is so susceptible to freezing, which renders it useless.

Luminar Alumina Coat is available in 1 quart, 1 gallon, and 5 gallon containers.

Fiber Molds

Mold Mix 6, used as a splash coat over a wax model before the less refractory plaster mold material is applied, lessens the possibility of surface cracking.

MOLD MIX 6 is a blend of graded sizes of ceramic fibers in an air-setting binder. It is a paste that becomes stiff upon losing its water, permanent at 500° F, and develops a ceramic bond when fired to fusing temperatures. It is packaged in quarts and gallons and is relatively inexpensive, when considering all its varied uses. It can be used as a glue for moldable fiber blanket, as a mold coat for rigidized fiber molds and as a castable mold material, by itself.

When used as a glue, it is thinned by adding 30% water. It is brushed onto both surfaces to be bonded, which are pressed together and fired to fusing temperatures. When Mold Mix 6 is used as a mold coat, it is mixed to a brushing consistency by adding 30% to 50% water. Two or three coats are brushed on, air drying between coats. It can be dried in a kiln or with a hot air blower, but should not be heated over 200°-250°F, or it becomes too hard to sand. Successive coats are added, sanding between applications, until the desired surface is acquired. Then it is fired to fusing temperature before shelf primer is applied.

Mold Mix 6 can be applied over any bas-relief model (pattern) after applying a parting agent. It is used in a water-thinned coat applied with a brush to pick up fine detail. The first coat is air dried for one hour or dried faster with hot air, then successive coats of approximately 1/4 to 3/8 inch, are applied, drying between applications. Mold Mix 6 is permeable and may be dried rapidly, without cracking or shrinking. If the pattern is not clay or wax, the microwave oven, set on medium, may be used for drying. After the pattern is removed from the dry mold, the mold is fired to 1300°F. This process can be used for small objects up to three or four inches in diameter. If no undercuts exist in the mold, many pieces may be taken from one mold. After two or three firings, apply Luminar Alumina Coat as a separating agent.

Hot glass can be delivered to this material when it is cold. Press molding or molten glass casting, covered in other chapters, are two of the ways of using Mold Mix 6 to its fullest potential. The porosity, thermal shock characteristics and good compressive strength make Mold Mix 6 the best (other than metal) mold material on the market that can be used in this way.

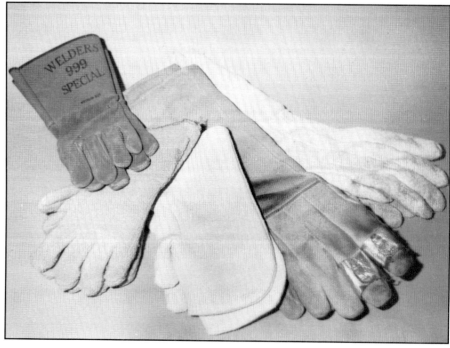

CHAPTER TEN

Safety

SAFETY

The intent of this section is to encourage awareness of potential hazards in the studio environment. As we become more knowledgeable about our environment, it becomes obvious that, as glass artists, we are surrounded by a variety of potentially hazardous materials and situations.

Common sense can play an important role in reducing or limiting dangerous exposure, and in maintaining a healthy studio environment. For example, materials that irritate the skin will invariably irritate the eyes and the respiratory tract. Materials such as paints and enamels, which contain lead and other metals, and which should not be inhaled in powdered form, may give off fumes when fired. The judicious use of gloves, safety glasses, and/or a respirator equipped with filters appropriate to the hazards involved will help minimize undesirable exposure. Confining use of hazardous materials to a particular area in the studio, good studio housekeeping habits, and the addition of exhaust ventilation localized to specific areas of the studio are other simple ways to minimize hazards.

PRODUCTS WHICH REQUIRE ATTENTION

SHELF PRIMER/KILN WASH contains silica (quartz) and is, therefore, hazardous to breathe in its dry form. Use localized ventilation or, better yet, work outdoors, when sanding shelf primer off of kiln shelves. Avoid inhaling the dust, as excessive exposure may cause silicosis. A respirator fitted with a filter approved for eliminating particulates should be worn when dust is created in an unventilated area.

CALCIUM CARBONATE is considered harmless but, in its raw form, may contain other minerals. Avoid inhaling dusts, since any dust may cause eye, nose and throat irritation.

SQUEEGEE OIL is a liquid whose combustible byproducts are toxic. When squeegee oil is fired, kiln ventilation is crucial, to avoid iritation to eyes and lungs. It is important to wash skin thoroughly after contact with the liquid oil.

OVERGLAZES usually contain material from the lead silicate family. Take care when using any product containing lead. Avoid breathing the dust or mist, and use localized exhaust ventilation. Do not smoke, eat, or drink while working with these materials; wash skin thoroughly after contact with overglazes.

FIBER MOLD HARDENER may contain ethylene glycol, considered highly toxic. Avoid inhalation, contact with the skin, and any availability to children who could accidentally ingest this material. Venting the kiln of fumes when firing, and wearing rubber gloves when working with this material will reduce exposure.

FIBER PRODUCTS contain aluminosilicate and may be irritating to the skin, eyes, and respiratory system, as well as being a possible cancer hazard. A dusty form of the fibers is released when the material is handled to cut or tear. In addition, all fiber materials leave a white, powdery residue on the glass and the kiln shelf when fired that, when disturbed, may release potentially hazardous dust into the air. A respirator equipped with filters to remove particulates should be worn when handling fired fiber; clean glass under water to reduce dust.

Opposite page: Safety is very important but I'd rather be dead than look like these nerds.

ENAMELS AND PAINTS contain lead and other toxic metals. Localized ventilation and/or the use of an approved respirator should be employed, whenever working with these materials in either dry or spray form. Tools and the work area should be kept clean and gloves should be worn, if you have cuts, burns, or other skin problems. Never eat or smoke while working with these materials.

Firing these materials can cause metals and opacifiers to fume; kilns should be properly vented to prevent contamination of the work area.

FRITS are usually of large enough grain size that they are not ordinarily inhaled. However, since there may be "fines" present, small enough to be inhaled and cause lung damage, use of a respirator is recommended when handling glass frit.

CASTABLE MOLD MATERIALS including plasters, investments, refractories, sand, and clay in the dry form should be considered hazardous, since most contain silica and may cause serious lung damage. Avoid creating a dusty atmosphere, wear a respirator, and work in a well-ventilated area. Localized exhaust is recommended.

USE OF STUDIO EQUIPMENT

KILNS-When kiln firing we not only employ volatile substances, in materials other than glass, that can be hazardous, but the heating of these substances produces a variety of gases and vapors. These include organic combustion products from paints and oils, as well as fumes from lead and other metals, fumes from mold hardener, and particulates from fiber products.

To help reduce exposure, it is advisable to install a ventilation system for the kiln. This can be a commercial or custom-built hood or a venting system designed to attach to, and ventilate, a kiln. Several of these are on the market and are readily adaptable to most kilns.

Protective clothing should be worn when opening a hot kiln. Electric shock will be prevented by turning the kiln off before reaching inside to comb or manipulate glass.

FURNACES/GLORYHOLES-It is important to make reference to local fire regulations and other codes that will affect the placement and use of furnaces and glory holes and the placement of propane tanks.

Bottled gases should be treated with respect; tanks should be secured and fittings well maintained. Where furnaces are fueled with natural gas piped to the burner, a periodic check of the fittings and joints with soapy water will reveal any leaks. Keeping gas plumbing in good repair will prevent explosions from escaped gas.

Safety glasses should be worn to protect your eyes, not only from flying glass shards, but from infrared light produced by the intense heat source. Prolonged exposure can cause cataracts.

Safety

Non-asbestos gloves, such as those made of Kevlar or Zetex, and clothing made of natural fibers can protect hands and arms during hot glass sessions.

An indoor glass furnace should be equipped with a localized ventilation system, to insure that the studio is not contaminated with fumes from the glass melt.

GRINDING AND POLISHING-Glass dust from "cold working" is hazardous. During wet grinding, glass dust accumulates as a paste and, when dry, becomes a hazardous powder. Equipment, clothing and work areas should be kept clean and a respirator worn, when necessary.

Common sense care should be taken around all "cold working" machinery, as with all power tools. Guards should be installed on grinding wheels and shafts that may catch hair or clothing. Safety glasses should be worn to protect the eyes from flying shards of glass.

Don't smoke and wear your glasses upside down. Mike Dupille, Ruth Brockman, Dan Ott and Boyce are 1) sometimes confused about safety, 2) will die of cancer and/or 3) are looking for trouble.

SAFETY EQUIPMENT

RESPIRATORS can be very effective in situations where specific localized exhaust ventilation is not possible. Respirators such as we use in our studio consist of a silicone mask worn over the nose and mouth, with a cylindrically-shaped, replaceable filter cartridge on either side of the face. Filters are designed for specific types of hazardous exposure.

It should be noted, however, that respirators do have drawbacks. In glass work we are exposed to particulates, fumes or both. Problems arise because fumes easily penetrate dust filters and dust quickly clogs fume filters. It is important to use the proper filter for the exposure, the best choice being a combination, a fume filter with a dust pre-filter. Respirators are not 100% effective and are not a substitute for good ventilation.

Respirators come in many sizes and shapes. Three companies that manufacture a variety of approved devices are:

Willson Co.
Division of WGM Safety Corp.
2nd and Washington Streets
Reading, PA

Survivair Division
U.S.D. Corp.
Santa Ana, CA

American Optical Corporation
P.O. Box 1
Southbridge, MA 01550

Check local safety supply outlets for additional information, or call manufacturers for names of distributors.

SAFETY GLASSES-Infrared wavelengths of light are produced by intense heat sources such as kilns, furnaces, and glory holes. Prolonged exposure to infrared light can cause cataracts. Glassworkers should always use protective glasses, when looking at infrared sources.

Calobar lenses block infrared, while transmitting visible wavelengths. This protects you from overexposure and still allows viewing of your work.

American Optical supplies data on three shades of Calobar lenses:

medium transmits:	52% visible light
	9% infrared
dark transmits:	33% visible light
	4.5% infrared
x-dark transmits:	17% visible light
	1% infrared

Didymium lenses should not be substituted for calobar lenses, as they filter only 20% of the infrared, not adequate protection. American Optical manufactures a variety of Calobar lenses which are available through local distributors.

GLOVES-Heat from equipment such as kilns, furnaces and glory holes, as well as from the glass itself can cause serious burns. Heat resistant gloves are a necessity for hot glass work and all activities involving a hot kiln. Gloves such as Zetex and Kevlar are available through most fusing and ceramic suppliers, and from:

Tempo Glove Mfg., Inc. A.R.T.CO
3820 W. Wisconsin Ave. 348 N. 15th St.
Milwaukee, Wisconsin 53208 San Jose, CA 95112

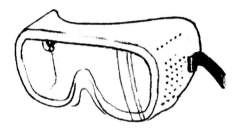

HOODS AND VENTS- Arranging good ventilation throughout the studio and localized ventilation, specific to problem areas, is the best way to avoid exposure to hazardous substances.

Use of a specially designed kiln ventilator is one way to achieve specific ventilation for fumes created in a hot kiln. There are several systems available commercially that can be added to most kilns:

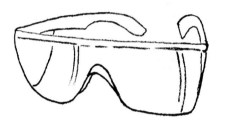

"Envirovent" "Direct Fume Vent System"
Skutt Ceramic Products Bailey Ceramic Supply
Portland, Oregon Kingston, New York

"Vent-A-Kiln System"
Vent-A-Kiln Corp."
Buffalo, New York

AN OVERVIEW

Awareness of such things as fumes, dust, and skin irritations and response to them with intelligent precautions will go a long way in preserving your health. *Pay close attention to warning labels and product safety information!* Some specific solutions for securing a healthy work environment are:

1. Use a respirator with filters approved for dust and fumes, when the situation suggests this precaution.
2. Install localized exhaust ventilation for specific problem areas.
3. Wear protective clothing (gloves, etc.) when working with irritants and other hazardous materials.
4. Keep the work area clean, vacuuming and mopping, as opposed to sweeping.
5. Lay out the studio so that problem areas are somewhat isolated and do not contaminate the entire studio. These areas should be easily ventilated.

MANUFACTURERS

CRUCIBLES

Cercon Ceramic Consultants, Inc.
P.O. Box 116
Hermann, MO 65041

Laclede-Christy Co.
P.O. Box 550
Owensville, MO 65066

ENAMELS, PAINTS AND LUSTRES

Drakenfeld Colors
P.O. Box 519
Washington, PA 15301

Englehard Corp.
Hanovia Hobby Products
Menlo Park, CN28
Edison, NJ 08818

L. Reusche & Co.
2-6 Lister Ave.
Newark, NJ 07105

Thompson Enamel
P. O. Box 310
Newport, KY 41072

FIBER PRODUCTS

Babcock & Wilcox Co.
Insulating Products Division
Augusta, GA 30903

Carborundum Resistant Materials Co.
Insulation Division
P.O. Box 808
Niagra Falls, NY 14302

C-E Refractories
P.O. Box 828
Valley Forge, PA 19482

Johns Manville Insulation
Drawer 17L
Denver, CO 80217

WRP
P.O. Box 2134
Elgin, IL 60120

Zircar
110 N. Main St.
Florida, NY 10921

GLASS CULLET, FRIT, FUSIBLE SHEET AND ROD

Bullseye Glass Co.
3722 S.E. 21st Ave.
Portland, OR 97202
(fusible sheet)

Corning Glass Works
Corning, NY 14830
(rods, tubing)

Glass Hues, Inc.
5650 Jason Lee Place
Sarasota, FL 33583
813-923-3334
(fusible sheet, frit, stringer)

Louie Glass Co., Inc.
Weston, WV 26452
(cullet)

CHAPTER THIRTEEN
Technical Information

Northstar Glassworks
9060 S.W. Sunstead Lane
Portland, OR 97225
(colored borosilicate)

Schott America/Desag
3 Odell Plaza
Yonkers, NY 10701
(fusable sheet)

Spruce Pine Batch Co.
Highway 19E
Spruce Pine, NC 28777
(glass batch)

Vitreous Group/Camp Colton
Colton, OR 97017
(glass stringer)

GLASS BLOWING TOOLS AND EQUIPMENT

A.R.T.CO
348 N. 15th St.
San Jose, CA 95112

Steinert Pipes and Rods
1000 Mogadore Rd.
Kent, Ohio 44240

GRINDING, POLISHING, AND SAWING EQUIPMENT

Amazing Glazing
2929 E. Coon Lake Rd.
Howell, MI 48843

Covington Engineering Corp.
715 West Colton Ave.
Redlands, CA 92373

C.R. Laurence Co., Inc.
2503 E. Vernon
Los Angeles, CA 90058

Crystalite Corporation
13449 Beach Ave.
Marina Del Rey, CA 90292

Gemstone Equipment Mfg. Co.
750 Easy St.
Simi Valley, CA 93065

Glastar Corp.
20721 Marilla St.
Chatsworth, CA 91311

Gryphon Corp.
101 E. Santa Anita Ave.
Burbank, CA 91502

Inland Craft Product Co.
32046 Edward
Madison Heights, MI 48071

MWP Associates
Box 99775
San Francisco, CA 94107

KILNS

Aerospex Company
1433 Roosevelt Ave.
National City, CA 92050

Denver Glass Machinery
3065 Umatilla St.
Englewood, CO 80110

Glass Glow
1122 Trade Center
Suite 500
Rancho Cordova, CA 95742

Jen Ken Kilns
4569 Samuel St.
Sarasota, FL 95841

Olympic Kilns
6301 Button Gwinnett Dr.
Atlanta, DA 30340

Paragon Industries
2011 S. Town East Blvd.
Mesquite, TX 75149

Seattle Pottery Supply
35 South Hanford
Seattle, WA 98134

Skutt Ceramics
2618 S.E. Steele St.
Portland, OR 97202

KILN CONTROLLERS

Advanced Technical Services, Inc
21045 Des Moines Memorial Dr.
Des Moines, WA 98198

Digitry Company
33 Ship Ave.
Medford, MA 02155

Edward Orton Jr. Ceramic Foundation
P.O. Box 460
Westerville, OH 43081

Kilntrol
8546 Madison Ave.
Fair Oaks, CA 95628

Paragon Industries
2011 S. Town East Blvd.
Mesquite, TX 75149

Seattle Pottery Supply
35 South Hanford
Seattle, WA 98134

MOLD MATERIALS

Perma Flex Mold Co.
1919 E. Livingston St.
Columbus, OH 43209

Ransom & Randolf
2337 Yates Ave.
Los Angeles, CA 90040

United States Gypsum
Tooling & Casting Division
101 S. Wacker Dr.
Chicago, IL 60606

RIGIDIZER

Carborundum Resistant Materials Co.
Insulation Division
P.O. Box 808
Niagra Falls, NY 14302

Zircar
110 N. Main St.
Florida, NY 10921

Technical Information

NEW RESOURCES

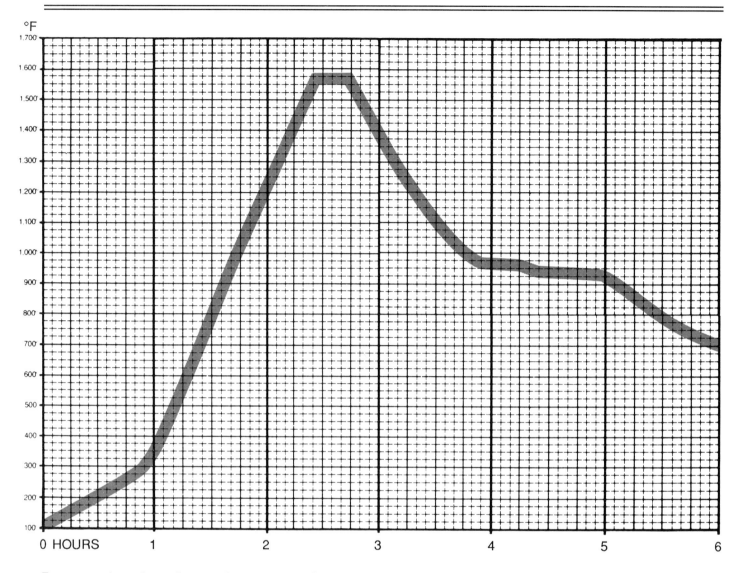

Temperature rise and annealing curve for 3/8" of glass.

FIRING AND ANNEALING

On these pages are graphs of the most commonly used firing and annealing procedures used in my classes. The first graph on pages 116 and 117 shows the initial heat cycle, the fuse soak, and the annealing schedule for fused work 1/4" to 3/8" thick and up to 12" across. When annealing glass pieces larger than 12 inches, I suggest adding ten minutes to the annealing time for each inch across over 12".

Temperatures shown on all graphs are pyrometer readings, not glass temperatures. The fact that the glass is cooler than the temperature reading on the pyrometer at the initial heat and hotter than the pyrometer indicates during cooling has been taken into consideration. The graphs are for 90 coefficient of expansion glasses.

The graph on page 118 shows annealing glass one-half inch thick and up to 12" across. Annealing time should be extended five minutes in each of the three annealing steps for each inch more than twelve inches across.

Technical Information

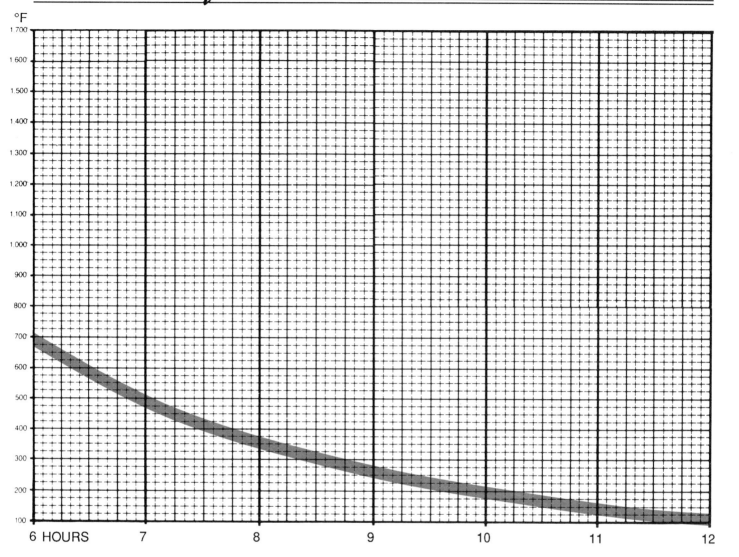

The graph on page 119 shows how to heat treat a piece of cast or fused glass up to two inches thick. This fast procedure works very well *if the object is fairly symmetrical, and does not change radically in thickness across the glass piece.* For example, a face or mask with a thickness of one-half inch at the edges, one inch at the cheeks, and two inches at the nose is commonly annealed in our classes using this schedule.

On pages 120 and 121 is a graph of the annealing time for a one inch glass slab, as suggested by Daniel Schwoerer in his technical paper at the end of this section.

When taking pieces of glass through the firing (heating) and annealing (cooling) stages there are three factors that should be kept in mind: the thickness of the piece or pieces of glass, the coefficient of expansion of the glass, and the size or volume of the finished object.

The initial heating stage consists of heating the glass from room temperature to above the strain point, about 850°F. During initial heating the outside of a piece of glass heats first and, as the molecules heat, they expand causing compression of the glass surface. If heating is not too rapid, the compression is balanced by tension in the middle of the glass piece. But if the thermal gradient is too large, the glass will break due to thermal shock.

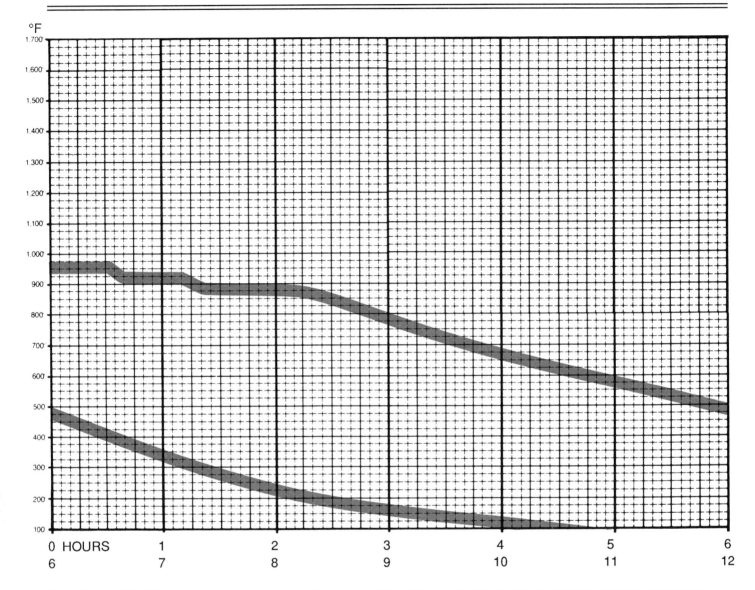

Annealing for 1/2" of glass.

I have found that it is prudent to divide the initial heating process into two stages: very slow rise to 350°F, then a somewhat more rapid rise to the strain point. Large pieces of glass, over 12" across, and tall stacks of glass, those four or more layers thick, should be heated at a rate of 7-10°F per minute. After this slow rise to approximately 350°F, a faster rate of 15-20°F per minute may be maintained up to the strain point. This rate of climb may be continued to the desired fusing temperature, or increased after reaching the strain point.

Heat soaking the glass at 1550°F (for Bullseye) for fifteen minutes gives more control of the surface line quality than does firing the kiln to a higher temperature for a shorter length of time. When the desired results are obtained, turn the kiln off and let it cool to the annealing temperature.

Technical Information

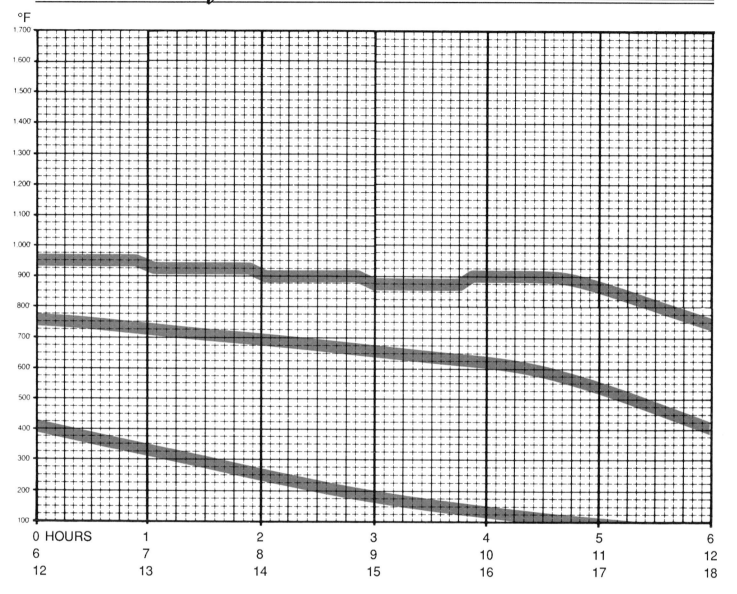

Heat treating for 1-1/2-2" cast glass.

    Top loading kilns or kilns with a symmetrical venting system may be vented one-half inch, in order to speed cooling. But if the kiln cannot be cooled symmetrically, it is best to let it cool at its natural rate, with doors and peep holes closed. Venting a kiln unsymmetrically will cause a thermal gradient within the glass *and* the kiln that will cause uneven annealing.

    The internal stress of any piece of glass depends on its thermal history, during cooling. A temperature gradient occurs whenever heat is conducted through any substance. As glass cools, the outside surface is always cooler than the inside. The thicker the glass and the larger the glass slab, the greater is this temperature difference. Heat soaking a slab of glass, within the annealing range, until the interior temperature is the same as the surface temperature, will remove the gradient, relieving the strain.

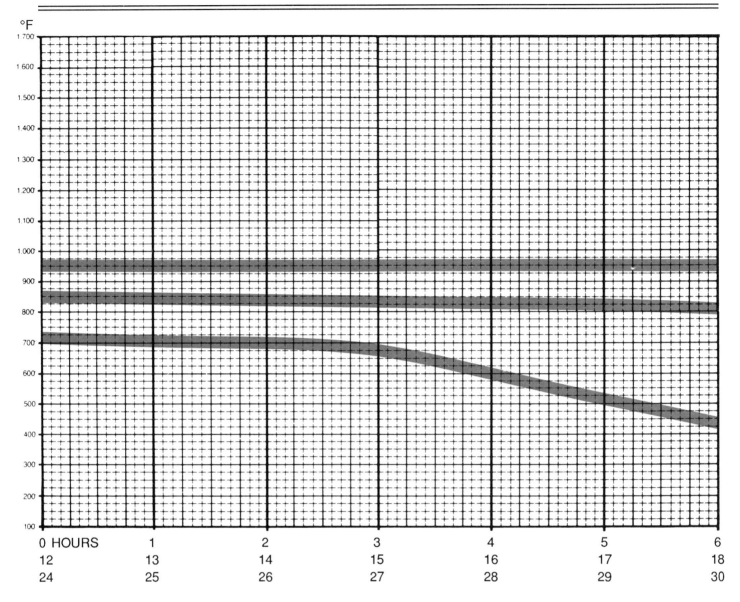

Graph shows theoretical annealing curve for 1" of glass

After heat soaking the glass at any temperature, it must be once again lowered in temperature, causing another thermal gradient. However, if most of the stress has been relieved by heat soaking and slow cooling through the annealing range to the strain point, it will be free of any permanent strain.

Any strain that occurs after the strain point will be temporary. If the cooling after the strain point proceeds at a uniform rate, there will be a constant temperature gradient between the surface of the glass and the middle of the glass. If the strain is not enough to break the glass during this cooling phase, from strain point to room temperature, and the entire slab becomes a uniform temperature, the temperature gradient disappears and so does the strain. As the glass slab approaches the size of the kiln shelf that supports it, the more likely it becomes that there will be a thermal gradient between the outside edges of the slab and the middle.

Technical Information

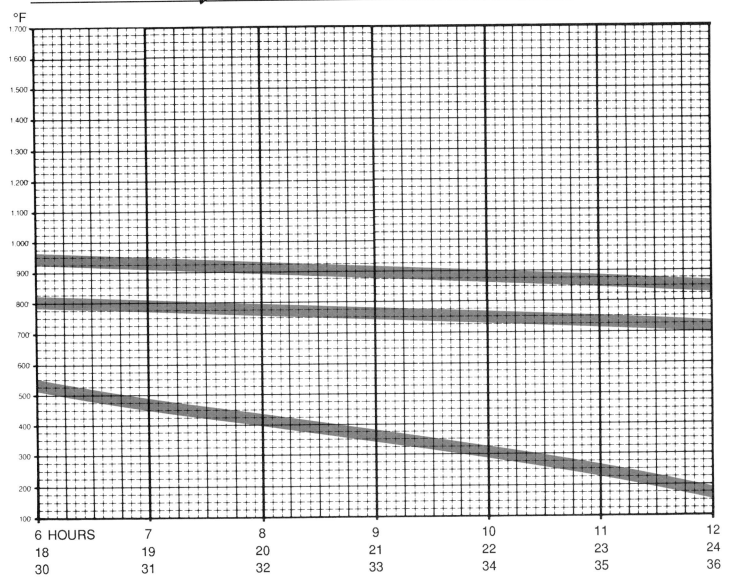

When these large pieces are annealed, heat must be added to the outside edges of the glass slab, not just over the top. For this reason it is important to have heating coils around the outside edges of the kiln shelf that supports the glass. In a side-fired kiln, the heat radiates from the side walls, where the elements are located, and goes toward the center of the kiln. The kiln shelf also conducts heat in this same direction, Therefore, the glass on the outer perimeter of the shelf receives heat before the glass in the center of the shelf does. Since the outside edges lose heat first during cooling, the ability to add heat to the edges during annealing is absolutely necessary when annealing glass slabs that are near the size of the kiln shelf, and as thick as one-half inch. Consequently, controlling heating elements independently for different areas of the kiln is critical when the glass fusing project is large and/or thick. The facility to have this control can, if necessary, be added to a kiln by installing a heating element just below the outer edge of the supporting kiln shelf.

TEMPERATURE CONVERSION TABLE

°C		°F	°C		°F	°C		°F	°C		°F	°C		°F
Find the temperature you want to convert, either °F or °C, in the center column of numbers. If the temperature is in °F, the figure to the left will be the correct temperature in °C; if the temperature is in °C, the figure to the right will be the correct termperature in °F.			20.0	68	154.4	138	**280**	536	421	**790**	1454	704	**1300**	2372
			20.6	69	156.2	143	**290**	554	427	**800**	1472	710	**1310**	2390
			21.1	70	158.0	149	**300**	572	432	**810**	1490	716	**1320**	2408
			21.7	71	159.8	154	**310**	590	438	**820**	1508	721	**1330**	2426
			22.2	72	161.6	160	**320**	608	443	**830**	1526	727	**1340**	2444
			22.8	73	163.4	166	**330**	626	449	**840**	1544	732	**1350**	2462
			23.3	74	165.2	171	**340**	644	454	**850**	1562	738	**1360**	2480
			23.9	75	167.0	177	**350**	662	460	**860**	1580	743	**1370**	2498
			24.4	76	168.8	182	**360**	680	466	**870**	1598	749	**1380**	2516
			25.0	77	170.6	188	**370**	698	471	**880**	1616	754	**1390**	3534
			25.6	78	172.4	193	**380**	716	477	**890**	1634	760	**1400**	2552
			26.1	79	174.2	199	**390**	734	482	**900**	1652	766	**1410**	2570
			26.7	80	176.0	204	**400**	752	488	**910**	1670	771	**1420**	2588
			27.2	81	177.8	210	**410**	770	493	**920**	1688	777	**1430**	2606
			27.8	82	179.6	216	**420**	788	499	**930**	1706	782	**1440**	2624
0	32	89.6	28.3	83	181.4	221	**430**	806	504	**940**	1724	788	**1450**	2642
0.56	33	91.4	28.9	84	183.2	227	**440**	824	510	**950**	1742	793	**1460**	2660
1.11	34	93.2	29.4	85	185.0	232	**450**	842	516	**960**	1760	799	**1470**	2678
1.67	35	95.0	30.0	86	186.8	238	**460**	860	521	**970**	1778	804	**1480**	2696
2.22	36	96.8	30.6	87	188.6	243	**470**	878	527	**980**	1796	810	**1490**	2714
2.78	37	98.6	31.1	88	190.4	249	**480**	896	532	**990**	1814	816	**1500**	2732
3.33	38	100.4	31.7	89	192.2	254	**490**	914	538	**1000**	1832	821	**1510**	2750
3.89	39	102.2	32.2	90	194.0	260	**500**	932	543	**1010**	1850	827	**1520**	2768
4.44	40	104.0	32.8	91	195.8	266	**510**	950	549	**1020**	1868	832	**1530**	2786
5.00	41	105.8	33.3	92	197.6	271	**520**	968	554	**1030**	1886	838	**1540**	2804
5.56	42	107.6	33.9	93	199.4	277	**530**	986	560	**1040**	1904	843	**1550**	2822
6.11	43	109.4	34.4	94	201.2	282	**540**	1004	566	**1050**	1922	849	**1560**	2840
6.67	44	111.2	35.0	95	203.0	288	**550**	1022	571	**1060**	1940	854	**1570**	2858
7.22	45	113.0	35.6	96	204.8	293	**560**	1040	577	**1070**	1958	860	**1580**	2876
7.78	46	114.8	36.1	97	206.6	299	**570**	1058	582	**1080**	1976	866	**1590**	2894
8.33	47	116.6	36.7	98	208.4	304	**580**	1076	588	**1090**	1994	871	**1600**	2912
8.89	48	118.4	37.2	99	210.2	310	**590**	1094	593	**1100**	2012	877	**1610**	2930
9.44	49	120.2	38	**100**	212	316	**600**	1112	599	**1110**	2030	882	**1620**	2948
10.0	50	122.0	43	**110**	230	321	**610**	1130	604	**1120**	2048	888	**1630**	2966
10.6	51	123.8	49	**120**	248	327	**620**	1148	610	**1130**	2066	893	**1640**	2984
11.1	52	125.6	54	**130**	266	332	**630**	1166	616	**1140**	2084	899	**1650**	3002
11.7	53	127.4	60	**140**	284	338	**640**	1184	621	**1150**	2102	904	**1660**	3020
12.2	54	129.2	66	**150**	302	343	**650**	1202	627	**1160**	2120	910	**1670**	3038
12.8	55	131.0	71	**160**	320	349	**660**	1220	632	**1170**	2138	916	**1680**	3056
13.3	56	132.8	77	**170**	338	354	**670**	1238	638	**1180**	2156	921	**1690**	3074
13.9	57	134.6	82	**180**	356	360	**680**	1256	643	**1190**	2174	927	**1700**	3092
14.4	58	136.4	88	**190**	374	366	**690**	1274	649	**1200**	2192	932	**1710**	3110
15.0	59	138.2	93	**200**	392	371	**700**	1292	654	**1210**	2210	938	**1720**	3128
15.6	60	140.0	99	**210**	410	377	**710**	1310	660	**1220**	2228	943	**1730**	3146
16.1	61	141.8	100	**212**	413	382	**720**	1328	666	**1230**	2246	949	**1740**	3164
16.7	62	143.6	104	**220**	428	388	**730**	1346	671	**1240**	2264	954	**1750**	3182
17.2	63	145.4	110	**230**	446	393	**740**	1364	677	**1250**	2282	960	**1760**	3200
17.8	64	147.2	116	**240**	464	399	**750**	1382	682	**1260**	2300	966	**1770**	3218
18.3	65	149.0	121	**250**	482	404	**760**	1400	688	**1270**	2318	971	**1780**	3236
18.9	66	150.8	127	**260**	500	410	**770**	1418	693	**1280**	2336	977	**1790**	3254
19.4	67	152.6	132	**270**	518	416	**780**	1436	699	**1290**	2354	982	**1800**	3272

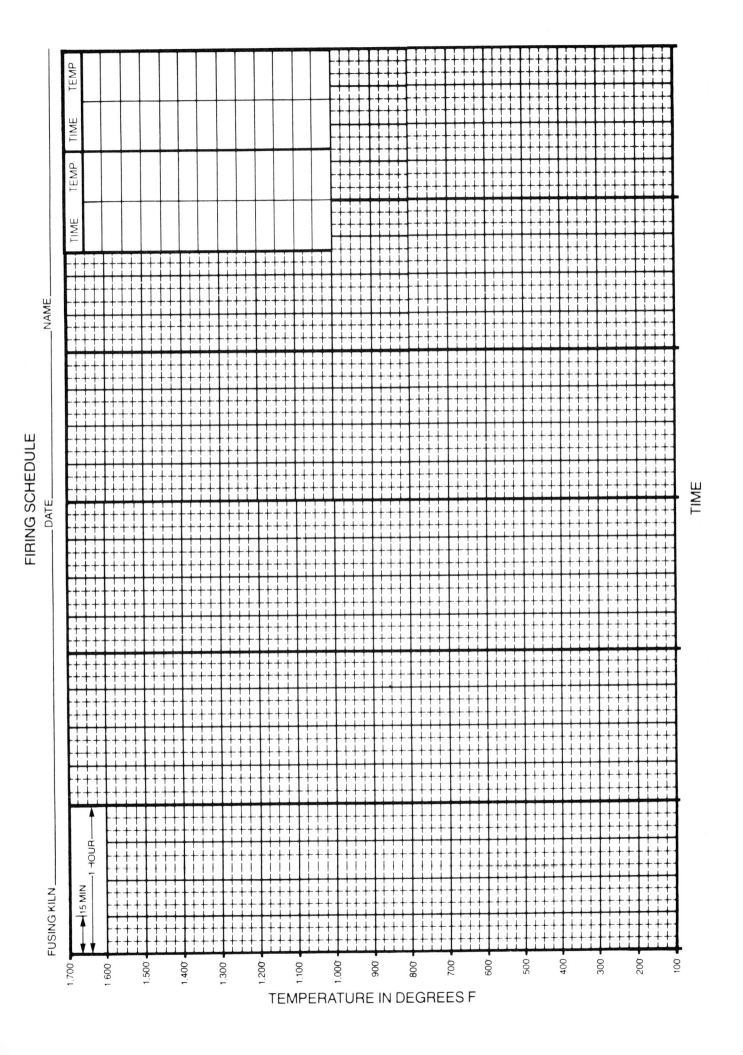

Technical Information

TECHNICAL ASPECTS OF FUSING WITH BULLSEYE GLASS

by Daniel Schwoerer
President, Bullseye Glass Company

KILNFORMED GLASS

Having co-authored a book a number of years ago that attempted to present some of the myriad possibilities available to anyone willing to reheat sheet glass, I hope now to be able to clarify and expand on some of the basic techniques presented there and to provide a little more information which will be of use to those who work with our glass in one of the numerous methods broadly categorized as "kilnforming". I feel also somewhat responsible to clear up a certain confusion of terminology resulting from publishing a book entitled *Glass Fusing Book 1* that addressed procedures rather broader in scope than the term "fusing" alone implies.

In its strictest definition glass fusing refers to the heat bonding of glasses. In its recent resurgence *fusing* has been loosely used to describe a wide variety of techniques involving the viscous manipulation of previously formed pieces of glass in a kiln. The glass may range from chunks of cullet to crushed, fritted or cut pieces of flat sheet glass to previously fused glass elements or forms. The procedures involving these materials may be done on flat kiln shelves, into or over molds. The degree to which the glasses are fused can vary from a low heat in which they are just tacked together, to a high heat in which they become a flowing liquid. Considering the diversity of techniques and materials involved, the term *kilnforming* would seem to more acurately describe this exceedingly broad area of glassworking.

FUSING

As a specific heat treatment of glass, fusing occurs over a broad temperature range with end results that are visually quite varied. At the lowest heat possible, fused glass pieces will stick together (the term "laminating" or "tack fusing" is frequently applied here), but retain almost all of their initial physical characteristics save a slight rounding of edges. At this heat level no noticeable flow or displacement of the individual pieces of glass occurs.

In fusing at higher heats, the separate pieces of glass completely lose their original shapes, flowing together, eliminating voids or spaces between them. Techniques at this heat level include indirect crucible pouring, direct mold casting, pate de verre etc.

Between these two extremes the possibilities for fusing pieces of preformed glass are virtually limitless.

COMPATIBILITY

To successfully fuse two or more glasses together all of the glasses need to be *compatible*. Compatibility is the characteristic of certain glasses that allows them to be fused together and — after proper cooling to room temperature — have no undue stresses that will lead to fracturing. The tendency of many colored glasses to be *incompatible* with each other has always been a major obstacle in fusing.

Compatibility requires that the various glasses expand and contract similarly upon heating or cooling. More technically stated, the glasses must have similar *coefficients of expansion.*. The coefficient of expansion is a number that expresses a percentage of change in length per degree of change in temperature. The coefficient of expansion is determined by measuring the change in length of glass for a one degree centigrade increase in temperature. (Fig 1). It is obviously a very small number (Bullseye glasses are in the vicinity of 0.0000090). For simplicity's sake when comparing expansion coefficients of different glasses, all the zeros are ignored. Bullseye glass is commonly referred to as having an expansion of approximately "90".

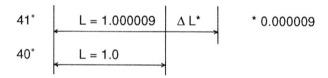

Fig. 1 Coefficient of Expansion = $\frac{\Delta L}{L}$ = $\frac{.000009}{1}$ = $90 \times 10^{-7} / C°$

The magnitude of the expansion number is a relative measure of how much a particular glass expands upon heating. The larger the number, the greater the expansion. Lower expansion glasses have better resistance to thermal shock. Pyrex (32.5), for example, will withstand a direct flame, while higher expansion glasses (those generally above 100) similar to many studio blowing glasses are more susceptible to shock and will not hold up well when subjected to temperature changes such as those experienced in repeated cycles of a dishwasher.

MEASUREMENTS OF EXPANSION

Unfortunately, most laboratory tests used to determine coefficients of expansion are performed over a temperature range from room temperature to 300°C (512° F). In fusing, the expansion from 300° C upward to the softening point is as important as the expansion in the lower range. In fact, the expansion above the strain point is usually 2 to 3 times greater than that below the strain point and in many cases this expansion rate is quite non-linear. This explains why laboratory determinations of expansions cannot necessarily be used as the sole criteria of compatibility for fusing.

Technical Information

Similarly, theoretical coefficients of expansion which are determined by mathematical formulas such as those of English and Turner can be extremely inaccurate when applied to formulas other than those of simple soda-lime glasses. Because calculations of this type are based on very specific ingredients and do not take into account variations in raw materials, melting schedules or the high percentages of various coloring oxides the results they yield when applied to complex glasses are frequently quite skewed.

For instance, English and Turner's calculations when applied to Bullseye's clear (#1101) and white opalescent (#0113) glasses yield coefficients of expansion of 84.5 and 76.5 respectively. The measured coefficients for these glasses are in fact 91 and 88. It is important to note that these glasses show no stress when tested for compatibility under conditions duplicating those of the actual fusing process.

CHIP TEST FOR COMPATIBILITY

The most relevant test for fusing compatibility anticipates and attempts to duplicate the same process that will be used in producing the final fused piece. Bullseye has employed this testing method for the last 9 years in generating its line of "Tested Compatible" glasses. The method is an adaptation of that used regularly in the technical glass community while incorporating the actual fusing process. I refer to it as the *chip test*.

Chips of colored glass approximately 1/2" square are fused to a larger base strip (we use strips 2" x 14") of clear glass. A control chip of the same base clear glass is also fused to the bar as a test of adequate annealing. (Annealing being the process of controlled cooling of the glass to avoid undesireable stresses in the final cold state.) After proper annealing the chips are viewed with a polariscope. Stress existing between the colored chip and the base clear glass will appear as a halo of light surrounding the chip. The intensity of the halo is a qualitative measure of the amount of stress. (See appendix A for more details). If little or no halo is evident, the glasses are compatible or "fit". All glasses which fit the same clear base glass will fit each other.

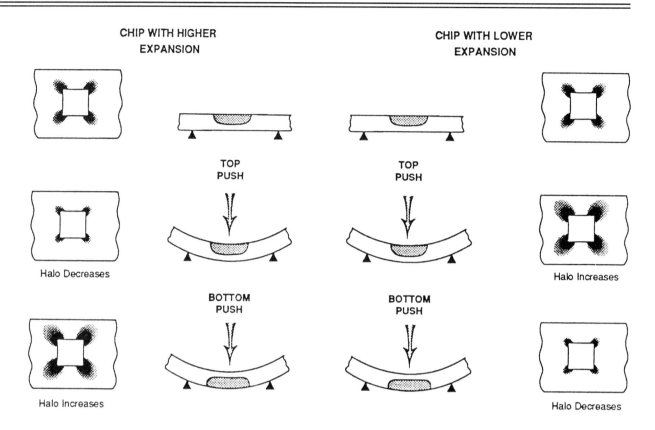

Fig.2: Procedure to determine whether chip glass is higher or lower in expansion than surrounding glass

In addition to compatibility information the chip test will reveal any color changes that may take place on re-firing of the glass, as well as the relative softening and flow characteristics. Additionally, a determination can be made as to whether the tested chip is higher or lower in expansion than the base clear glass by applying downward pressure on the chip and observing the change in intensity of the halo as described in Figure 2 above.

At Bullseye we have for a number of years accepted, marked and sold as "Tested Compatible" only those glasses which showed *no halo at all* in our factory compatibility testing. This does not mean that glasses showing slight stress or haloing will not work perfectly adequately for most fusing applications. By maintaining standards higher than are generally necessary we have tried to safeguard against any slight variations that might occur over time both in our own production or in the studio fuser's use of our glass. We have attempted to maintain a standardized clear glass which could be used for testing of all our other glasses and which would remain constant in succeeding years, insuring compatibility not only between sheets and runs, but year after year. I believe our success in accomplishing this goal is attested to by the number of artists who have regularly and successfully used our glasses in kilnforming applications over the last 9 years.

Technical Information

TOLERANCES OF STRESS

If, however, a *total* absence of haloing or stress in compatibility testing is not necessary for most successful fusing procedures, what are in fact the allowable tolerances?

Acceptable levels of stress whether related to compatibility or annealing should be determined by consideration of the following:

• *The size & thickness of the finished piece.*

The tolerances for jewelry work will clearly be much greater than those acceptable when fusing multi-layered constructions 20" in diameter.

• *Any secondary processing which will later be done to the work.*

If the work is to be sandblasted, ground, carved or otherwise subjected to surface abrasion it will be less able to tolerate stress.

• *The object's intended use.*

Objects which will be used in a functional way, subjected to frequent handling or repeated temperature changes will need to be freer of stress than those destined for the gallery or museum collection.

ANNEALING

The successful annealing of Bullseye's glasses, like the successful annealing of all glasses, is largely dependant upon the 3 points noted above. Annealing, however, —unlike compatibility — is affected by a 4th variable: the kilnforming procedures and conditions which precede and/or accompany it. For instance, if the kilnforming process involves uneven heating or cooling of the glass surface such as occurs in crash cooling the kiln to "freeze a shape" the glass may need to be soaked at a higher heat before proceeding with the annealing schedule. Or if the glass has been formed in a massive mold the heat transfer between the mold and the glass may require altering the annealing schedule.

Annealing schedules should be determined not only by the glass, its composition, size, end-use, and cold-working eventualities, but also by the size and configuration of the kiln or lehr in which the controlled cooling will occur.

In order to determine a proper annealing cycle certain technical data relating to the glass is necessary. These are: the softening, annealing, and strain points of the glass. For two typical Bullseye glasses the relevant technical data (and coefficients of expansion) are as follows:

	1101F Clear	**0113F White opalescent**
Softening point:	1260°F	1270°F
Annealing point:	990°	935°
Strain point:	920°	865°
Measured coefficient of expansion (0-300°C):	91	88

The information above enables us to compute an accurate annealing cycle by indicating the rate of release of stress at various temperatures in the annealing region.

It is difficult for stress to exist above the softening point since the glass quickly flows to relieve it. As glass cools it becomes stiffer and hence cannot relieve stress as easily by flowing. At the annealing point any stresses existing in the glass can be relieved in minutes. At the strain point, however, it will take hours to relieve the same stresses. *Below* the strain point stresses in the glass will not be relieved over any length of time.

From this it would appear best to soak at the annealing point for a short time and cool slowly down through the strain point to room temperature.

In practice, however, it is best to anneal by soaking the glass for a much longer time somewhere well below the annealing point but above the strain point, cool very slowly to some temperature well below the strain point and then cool to room temperature.

Recommended annealing schedules for various thicknesses of Bullseye glasses are given in table 1. The table was developed directly from information available in *The Glass Engineering Handbook*, 3rd Edition by McLellan and Shand (Chapter 4, "Stress Release and Annealing")

Thickness	Soak Temp	Soak Time	First Cooling Stage (Maximum)	2nd Cooling Stage (Maximum)	Final Cooling to Room Temp (Max)	Total Time
1/8"	930°F	1 Hr	7°/min to 855°F	14°/min to 765°F	20°/min	1 Hr 50 min
1/4"	930°	2 Hrs	2°/min to 845°	4°/min to 755°	20°/min	3 Hrs 39 min
1/2"	930°	4 Hrs	0.5°/min to 825°	1°/min to 735°	5°/min	11 Hrs 12 min
1"	930°	8 Hrs	0.15°/min to 785°	0.3°/min to 695°	1°/min	39 Hrs 27 min

Table 1: Annealing Schedules for Bullseye Glass

It is necessary to keep in mind that the annealing schedules given in Table 1 are what the *glass* should experience. It is necessary to take into account the difference between the glass temperature and kiln temperature during firing. As a general rule (with variations specific to each kiln and firing schedule) the temperature of the glass will be 50°F lower than the air temperature of the kiln on heat up and 50°F hotter than the air temperature on cool down.

In the final analysis there are no pre-determined schedules which can take into account all the variables involved when firing Bullseye or any other glass. However, given the information detailed above, an awareness of his/her own equipment, and a systematic approach to observation and record-keeping, the kilnworker should have little difficulty in successfully fusing and annealing Bullseye.

To quote Paul Marioni: "Annealing formulas are like bread recipes — everyone has one."

Technical Information

Appendix A
MECHANICS OF CHIP STRESS TEST

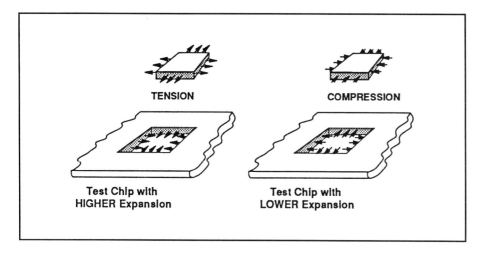

A chip of colored glass with a higher expansion will shrink **more** than the surrounding clear glass. This will cause both the colored chip as well as the surrounding clear glass to be in tension at the interface of each.

Likewise, a chip of colored glass with a lower expansion will shrink **less** than the surrounding clear glass, creating compression at the interface of each.

The tension or compression will cause halos of light to appear at the interface of the chip and the base glass when viewed with a polariscope.

The test appears to be more sensitive to measurements of tension than those of compression. The accuracy of the test is probably + 1 expansion point when the chip is of a higher expansion and + 2 points when the chip is of a lower expansion.

For further information refer to *The Handbook of Glass Manufacture Vol II* by Dr. Fay V. Tooley, Section 14 "Annealing and Tempering".

Technical Information

ANNOTATED RESOURCE INDEX

Anderson, Harriette, *Kiln Fired Glass*, Chilton Books, Ontario, 1970

    Out of print, but still fairly easy to find in used book stores and many libraries, this book contains many useful ideas and possible directions for the hobbyist. Anderson writes of her personal experiences with the use of enamels.

Arwas, Victor, *Glass-Art Nouveau-Art Deco*, Abrams, Inc., New York, 1987

    This is an historical book covering the background of glass artists and their techniques from the late 1800's to 1940. Daum, Galle, Lalique, Marinot, Tiffany, and others are discussed in detail. It contains over 250 color pictures, including photos of works from private collections not usually seen in other publications.

Bloch-Dermant, *The Art of French Glass*, Viking, New York, 1974

    This book is similar to the Arwas book, above, but not as broad in scope. Excellent photographs, 118 of which are in color. This book is full of detailed and technical information on France's leading artist-glassmakers. It is written simply and directly. The last chapter on pate de verre offers historical and technical insights into this glass process.

Cummings, Keith, *The Technique of Glass Forming*, B.T. Batsford, LTD, London, 1980

    Out of print, but worth a look, if you can find it, Cummings' was the first visually stimulating book on kiln forming glass. The section on press forming glass with a wooden mold, and then slumping over a form, is complete and makes the process inviting. Cummings covers glass art history, providing photos of the work, followed by an expose of how, in his view, the glass pieces were made.

Denoel, *La Pate De Verre*, Editions Denoel, Paris, 1984

    This book is a visual feast of traditional French pate de verre work. Although I have not read a translation of the French text, I find the work put together in this book worthwhile. Many of the photos are of work not seen in other publications, and 40% of the photos are in color.

Duthie, Arthur, *Decorative Glass Processes*, Corning Museum/Dover, New York, 1982

    Originally published in 1911, this book was the first comprehensive volume on flat glass decorative processes that was written in English. It was reprinted by the Corning Museum of Glass in appreciation of its excellence. It is amazing to read a book by a skilled craftsman writing 80 years ago, who seems as you read his work, to be a contemporary. The sections on patents and "recent" developments of sandblasting and electroplating are both interesting and comical.

Glasses, Borax Consolidated Limited, London, 1965
      This small book was written by Borax Consolidated Ltd. to explain how boric oxide is used in the composition of a wide range of glasses. It explains glass forming technology in a way that a novice can understand. Batch recipes and glass materials are listed. Melting procedures are described simply. The methods of forming glass containers, fiberglass, and technical glasses are explained with excellent illustrations. This is absolutely the best written book on industrial glass chemistry and forming methods available. This book was written as a sales tool and information guide for the glass industry. It can be obtained by writing to Borax Consolidated Limited, Borax House, Carlisle Place, London, England, SW1P 1HT

Hodkin, F.W., *A Textbook of Glass Technology*, Van Nostrand Reinhold, New York, 1927
      This is a reference book on the chemistry of glass, and industrial forming techniques. Although it has been out of print for sixty years, it can still be found in some reference libraries.

Goldstein, Sidney, *Pre-Roman and Early Roman Glass in The Corning Museum of Glass*, Corning Museum, Corning, New York, 1979
      There is no better resource for insight into the forming methods used by the Romans than the first ten pages of this book. The balance of the book is a picture reference guide to glass in the Corning Museum.

Illustrated Science and Invention Encyclopedia, H.S. Stuttman Co., New York, 1977
      The section on glass has the most complete illustrations and explanation that I have found on the float glass forming process.

Lalique Glass, Corning Museum/ Dover, New York, 1981
      This is an excellent pictorial survey of this most important glass factory. Looking at the accomplishments of Lalique makes one feel rather insignificant, but also challenged. Included are photos of some of the best cast crystal glass work I have seen.

Le Blanc, Raymond, *Gold Leaf Techniques*, Third edition, St. Publications, Cincinnati, Ohio, 1986
      This was written for sign painters, and is the most complete text available on the subject of gold leaf. This book is a great help to the glass artist who wants to use gold in conjunction with glass.

Lundstrom and Schwoerer, *Glass Fusing Book One,* Vitreous Publications, Portland, Oregon, 1983
      Now here's a book! This is still the most accurate and comprehensive material available on basic glass fusing.

Masterpieces of Glass, Corning Museum/Abrams, Inc, New York, 1980
      This is an excellent historical survey in one volume.

Technical Information

Pfaender, Heinz, *Schott Guide to Glass*, Van Nostrand Reinhold, New York, 1983

    This book explains the fundamentals of glass: what it is made of, how it is made, and how it has been used. It is a complete reference on all modern applications of glass processes and uses, and includes information on obscure topics such as sealing glass, conductive glass, foam glass, and potential future uses of glass.

Reynolds, Gil, *The Fused Glass Handbook*, Fusion Headquarters, Portland, Oregon, 1987

    As a book directed to the hobbyist, it has simple, easy to follow directions.

Scholes, Samuel, *Modern Glass Practice*, Cahners Publishing, Boston, 1975

    This technical book is the number two reference, in my regard, on glass composition and melting. Written as a text book for students in glass technology, it begins with the elementary physics of the glassy state, then continues through the chemistry of glass color and composition, mixing batch, furnaces, melting, and annealing.

Schuler, Frederick and Lilli, *Glassforming*, Chilton, Philadelphia, 1970

    This work covers flame working, enamels, glass color, and mold forming of glass in an easy to understand way. The reader may wish there were more information on each process. The technical information on glass is very complete, but the forming techniques and the examples used are very basic.

Schultes and Davis, *The Glass Flowers at Harvard*, E.P. Dutton, New York, 1982

    The flowers in this collection are unbelievable and, as a reference, the book provides an idea of what kind of detail can be accomplished with the lampworking process.

Shand, E.B., *Glass Engineering Handbook*, McGraw-Hill, New York, 1958

    This is the number one reference book on glass formulation, melting, and annealing. It is an excellent reference, but may put you to sleep.

Vargin V.V., *Technology of Enamels*, Hart, Inc., New York

    Translated from the Russian, this is the most complete book on enamel technology that I know of. It can be hard to understand unless you have a background in chemistry. It gives excellent information on making enamels fit other materials. Although the book is out of print, it can be found in reference libraries.

Weyl, W.A., *Coloured Glasses*, Sheffield, 1951

    This is a very technical book, the best around on coloring glass. To benefit fully from its contents, one needs to have a masters degree in glass chemistry, but the novice glass melter may get some good ideas about glass colors.

Education at Camp Colton

These days we seem to talk about The Glass Program at Camp Colton. Yet the camp had a long and illustrious history before it was ever invaded by anyone with glass madness. Before we bought the camp, it had a 55 year history as a Lutheran children's camp, also used for 4-H and outdoor school. Anyone who experienced its delights never forgot the name or the place that brought them such inspiration. I was one of those who was so affected. And so were my grandfather, my father and my son.

When I was old enough to travel to Oregon from California to visit my grandparents in Colton, my granddad frequently took me to the camp to experience the wonders there that delighted him. He and the other early members of Colton Lutheran Church had toiled to make the beautiful place by the creeks into a strong organization camp.

For the years between 1980 and 1984 I was a traveling minister of glass fusing. I seemed to be asked once every few weeks to travel to Maryland or Georgia or Australia or Japan to tell people how to fuse glass. When Kathy and I bought Camp Colton in 1985, to save it from the destruction that would have been inevitable at the hands of those who saw it as ready for development as real estate, we felt it would take nearly every minute at home (at camp) to find ourselves equal to the task of bringing the camp back to usefulness. We decided to find out if people would like to come to western Oregon and stay in one its most beautiful settings, while learning all the aspects of the craft of glass fusing. And so began The Glass Program at Camp Colton.

Kathy has occasionally gotten the award for most successful "firings" during a session.

Since that decision was made, we have been enlightened and exhausted and delighted by a continuing stream of student guests from all over the world. Our classes are held at various times of the year, but not at all times of the year. We have never stopped being amazed at the diversity within the groups we host. The individuals vary widely in age—from 22 to 80, with an average age in the mid-forties. They are equally diverse in background and degree of experience with glass. And I find it hard to think of any one of them who didn't have something special to give the group that convened here, and to us! Our sessions are quite small, and each group seems to become a family during the stay.

The Glass Program at Camp Colton is designed to bring students together in a situation where they can discover the materials and equipment available, learn the technology and sources for these, and prepare themselves to use that information to implement their own goals for designing with glass. Classes are outlined and structured so that a large body of knowledge can be covered. The teachers are people who love sharing a knowledge base and know how to communicate. The Glass Program at Camp Colton strives to provide a solid background in kiln fired glass, which can be built upon by the individual.

Camp Colton is thirty-five miles southeast of Portland, Oregon, in the Mt. Hood foothills. It consists of fifty-six acres of beautifully wooded land with two ponds (where I raise trout) and two streams. The trees are Douglas firs, some over 200 feet tall, and Western red cedar, mixed with a pleasant smattering of deciduous Alders and Maples. There seems to be about the place an essence that speaks, even to newcomers, of its history and the integrity of its founders. All this makes for an unusually pleasant setting, no doubt about it. But it is also freeing and serene and the students respond accordingly.

Boyce watches a six-pound trout follow his fly.

During their stay our student guests occupy small, sparsely furnished rooms, in the sunny, open area next to the lake. They respond to the old dinner bell for their meals, served in Riverfalls Lodge, located at the juncture of the two creeks. The glass seems to tempt them to spend long hours in the studios, but we manage to lure them outdoors for hikes, trout fishing, bowling ball croquet, evening campfires, and frequent pre-dinner horseshoe games.

Our students discover that we have reason to be as proud of our meal service as we are of the depth of the glass education we provide. We have focused with success on serving unusual, healthful, attractive menus. They are carefully prepared from natural foods, include homemade breads and desserts, and emphasize local foods including, of course, our own trout.

While lecture hours are specific, individual work proceeds at any pace that suits the individual. With the particular combination of facilities, the small size of our groups, the style of our meals, and the flexibility of the class work, we find we have been able to make students of all ages and levels of experience and energy comfortable. The goal, after all, is to help those individuals pursue the knowledge base they need to advance in their work. And we know that the glass education available at Camp Colton is well designed to do just that.

About the Author

I began working with glass in 1965, when I joined the new glass program established by Dr. Robert Fritz that year at San Jose State University in San Jose, California. At that time I was a ceramics major, studying with one of the great ceramic glaze technicians of our time, Dr. Herbert Sanders. The close correlation between the calculating and making of ceramic glazes and the process of making glass is a natural one. So, as a potter studying glaze calculation, I found it natural to apply the technology to glass, and was soon drawn by the material.

Dr. Fritz introduced glass and glass blowing as a medium for art. He put special emphasis on learning the nature and behavior of different glasses. His program matured over three years to the point that each individual student could learn to control all phases of the process of making finished blown objects. We built glass melting equipment, calculated and melted batch, formed the glass, and carried out all the cold working processes for finishing the annealed work. This, my introduction to glass, has never been forgotten, and my desire since then has been to share as much with others interested in glass as Robert Fritz did with me.

After my graduation from San Jose State, in 1967, I set up a ceramics and glass studio in southern California, teaching and selling my work there for two years until my wife and I moved to Corvallis, Oregon, in late 1969. At that time, in fulfillment of alternative service for the United States Government, I set up a ceramics and glass program for the Children's Farm Home, near Corvallis.

At the end of two years alternative service, I organized a home blowing studio in Corvallis, where I blew glass for galleries and craft fairs. While participating in craft fairs and shows, I met many other glass artists who had become infatuated with hot glass in the early years of the studio blowing movement in this country. We were all struggling to support our individual studios and families, while experimenting with new glasses and equipment.

Two of those artists were Ray Ahlgren and Dan Schwoerer, who were partners in a glass blowing studio in Portland, Oregon. It seemed that their experience and love for glass were similar to my own. Our experience in the glass world pointed to a need for more colored sheet glass for the stained glass industry. Forming a partnership in 1974, we established Bullseye Glass Company, the first new glass manufacturer to produce opal sheet glass since 1900.

For the next four years the pressing demands of an infant company consumed all of my time. As the company president in charge of administration and sales, I had little time for creative work. However, in 1978 I began designing independent stained glass panels, executed for me by more capable craftspersons. Although I completed a relatively large body of work, I was unsatisfied with the black lines and the cartoon effect created by the lead and copper foil. Even though I had control of the colors and texture of all the glass I used (I could make my own glass in the factory), I was not happy with the results.

I had met Kay Kinney during my years in southern California, and was aware of her struggles with fusing and laminating glass. In books I saw ancient Egyptian fusing, as well as fused work by contemporary artists Michael and Frances Higgins and Maurice Heaton. Since, at Bullseye, we produced mixed colors of glass daily, and had control of the formulas, it seemed a foregone conclusion that we could make sheet glass with similar coefficients of expansion.

The thought process went something like this: if sheet glasses had the same coefficient of expansion, they could be cut into shapes and fused together and there would be no need for all those cartoon lines. So, I started experimenting in 1979 or 1980—I don't know exactly when because the process was slow at first, fraught with many failures and just a few successes. If there was one memorable breakthrough, it was the application of the method of testing for stress with a polarimeter (from glass blowing) to glasses fused to a clear sheet glass with a constant coefficient of expansion.

When making sheet glass it is not important to have a constant coefficient of expansion among all the glasses. Single colors can all be different and mixed colors only have to be within one or two coefficient points of one another. In glass blowing it is not uncommon to use glasses together that vary in coefficient of expansion by four or five points, because the casing process holds the glass together. But when fusing glass flat, the glasses must be very close in coefficients. Establishing a clear glass as a constant, and then formulating the melt for all colors to fit that constant, made the contemporary glass fusing movement possible.

The ability to fuse glass, by taking it through the complete process of heating, holding and annealing, then checking the finished results with an accurate test, really stimulated my dreams of unlimited possibilities. I saw kiln fired glass as the wave of the future, providing freedom for all those who would like to be freed of the lead lines! Tiles, windows, bowls, sculptures, and building faces could all be made with fused sheet glass.

By 1981 I became adamant about producing glass for the fusing market at Bullseye Glass. My remaining partner, Dan Schwoerer, supported me in my one-man campaign to make fusing available to everyone, and during the next few years we succeeded in making available a line of fusing compatible glasses, teaching fusing in diverse parts of the world, establishing a line of products, besides glass, that fusers needed for their work, working with kiln manufacturers to get kilns designed for glass on the market, and writing *Glass Fusing Book One*.

The market was slow to move toward my vision for the fusing movement, and all that energy took its toll in a company whose focus remained the production of quality hand rolled sheet glass. In 1985, at the same time that Camp Colton came into our lives, I sold my shares of Bullseye to Dan, who keeps the company on track, making the best sheet glass they can make, leaving me free to concentrate on those questions that concern glass fusing.

Whereas the material glass continues to fascinate me, so much of what I do is stimulated by other people. My students and my friends in the glass art world have always challenged me with their questions and provided help in testing my theories.

NOTES

NOTES

PUBLICATIONS ORDER FORM

GLASS FUSING BOOK ONE	History and Contemporary Work • Basic Fusing Techniques • Compatibility • Firing and Annealing • Tools & Kilns • Molds • Slumping • Finishing Processes	_____ @ $30.00 = _____
PROJECTS IN KILN FIRED GLASS	A Set of Four Project Packets, Including Patterns, which Teach Specific Fusing Lessons: Wire Inclusions • Use of Glue • Stacking Methods • Production Work • Slumping • Pattern Placement	_____ @ $10.00 = _____
ADVANCED FUSING TECHNIQUES Glass Fusing Book Two	Bas Relief • Fiber Paper • Glory Hole • Sagging and Slumping • Designing • Using Stringer • Enamels and Lusters • Iridizing • Safety	_____ @ $40.00 = _____
GLASS CASTING AND MOLDMAKING Glass Fusing Book Three	Pate de Verre • Metal, Clay and Sand Molds • Kiln Forming in Sand • Lost Wax Casting • Castable Mold Materials and Formulas • Crucible Furnaces • Fiber Molds	_____ @ $40.00 = _____
SHIPPING, DOMESTIC, EACH BOOK OR PROJECT SET	Shipping will be U.P.S. in the states, unless otherwise arranged. Out of country buyers should inquire for shipping charges	_____ @ $2.50 = _____

TOTAL CHECK ENCLOSED: $ _____

Photocopy, fill out, and send with your check to:
**VITREOUS GROUP/CAMP COLTON
CAMP COLTON
COLTON, OR 97017
(503) 824-3150**